Introduction to continuum damage mechanics

MECHANICS OF ELASTIC STABILITY
Editors: H.H.E. Leipholz and G.Æ. Oravas

H.H.E. Leipholz, Theory of elasticity. 1974.
ISBN 90-286-0193-7
L. Librescu, Elastostatics and kinetics of anisotropic and heterogeneous shell-type structures. 1975.
ISBN 90-286-0035-3
C.L. Dym, Stability theory and its application to structural mechanics 1974.
ISBN 90-286-0094-9
K. Huseyin, Nonlinear theory of elastic stability. 1975.
ISBN 90-286-0344-1
H.H.E. Leipholz, Direct variational methods and eigenvalue problems in engineering. 1977.
ISBN 90-286-0106-6
K. Huseyin, Vibrations and stability of multiple parameter systems. 1978
ISBN 90-286-0136-8
H.H.E. Leipholz, Stability of elastic systems. 1980.
ISBN 90-286-0050-7
V.V. Bolotin, Random vibrations of elastic systems. 1984.
ISBN 90-247-2981-5
D. Bushnell, Computerized buckling analysis of shells. 1985.
ISBN 90-247-3099-6
L.M. Kachanov, Introduction to continuum damage mechanics. 1986.
ISBN 90-247-3319-7
H.H.E. Leipholz and M. Abdel-Rohman, Control of structures. 1986.
ISBN 90-247-3321-9

Introduction to continuum damage mechanics

by

L.M. Kachanov

Brookline, MA 02146
USA

1986 SPRINGER-SCIENCE+BUSINESS MEDIA, B.V.

Library of Congress Cataloging in Publication Data

Kachanov, L. M. (Lazar' Markovich)
Introduction to continuum damage mechanics.

(Mechanics of elastic stability ; 10)
Bibliography: p.
Includes index.
1. Continuum damage mechanics. I. Title.
II. Series.
TA409.K29 1986 620.1'126 86-5368

DOI 10.1007/978-94-017-1957-5

Contents

Chapter 3 - Creep and Fracture under Multiaxial Stress

Chapter 4 - Crack Growth under Creep Conditions

Chapter 5 - Damage Model for Ductile Fracture

Chapter 6 - Fatigue Damage

PREFACE

Modern engineering materials subjected to unfavorable mechanical and environmental conditions decrease in strength due to the accumulation of microstructural changes. For example, considering damage in metals we can mention creep damage, ductile plastic damage, embrittlement of steels and fatigue damage.

To properly estimate the value of damage when designing reliable structures it is necessary to formulate the damage phenomenon in terms of mechanics. Then it is possible to analyse various engineering problems using analytical and computational techniques.

During the last two decades the basic principles of continuum damage mechanics were formulated and some special problems were solved. Many scientific papers were published and several conferences on damage mechanics took place.

Now continuum damage mechanics is rapidly developing branch of fracture mechanics.

This book is probably the first one on the subject; it contains a systematic description of the basic aspects of damage mechanics and some of its applications.

In general, a theoretical description of damage can be rather complicated. The experiments in this field are difficult (especially under multiaxial stress and non-proportional loading). Therefore, experimental data, as a rule, are scarce. Determination of functions and constants, which play a role in the complex variants of the theory, from available experimental data is often practically impossible.

L.M. Kachanov

The problems of damage mechanics are mainly engineering ones. Therefore, the author tries to avoid superfluous mathematical formalism.

Some more details of the book's subject can be found in the list of contents.

The author wishes to express his thanks to Professor H.H.E. Leipholz for editing the manuscript and for many valuable remarks.

Chapter 1

INTRODUCTION

1.1 Some Types of Damage

Modern engineering materials subjected to unfavorable mechanical and environmental conditions undergo microstructural changes which decrease their strength.

Since these changes impair the mechanical properties of these materials, the term damage is used.

Consider some examples of damage of materials (mostly metals).

(a) *Creep Damage.* At high temperatures and under action of stress an accumulation and growth of microvoids in metal grains takes place (ductile transgranular creep fracture). At the same time there occurs an accumulation and growth of microcracks on intergranular boundaries (brittle intergranular creep fracture).

So, two different mechanisms of fracture take place simultaneously.

(b) *Ductile Plastic Damage.* The same phenomenon, i.e., nucleation and growth of microvoids and microcracks, takes place in metals as the result of a large plastic strain. This process leads to the so called plastic fracture.

(c) *Fatigue Damage.* Under the action of cyclic loading a gradual deterioration of the structure of a material, caused by the accumulation and growth of micro and macro cracks, takes place.

(d) *Embrittlement of Steels.* Under the action of atomic radiation the structure of the steel is changed, which decreases its plasticity and leads to its embrittlement.

As a result of the contact of steel with free hydrogen the atoms of the latter diffuse into the atomic grid of the steel, which leads to dangerous embrittlement of the structure ("hydrogen brittleness").

(e) *Chemomechanical Damage.* Under the action of a tensile stress (especially – the cyclic one) metals operating in agressive media (for example, in sea water) are subjected to intensive corrosion ("stress corrosion") or to some other chemical reaction.

(f) *Environmental Degradation.* Some materials (geo-materials, polymers,...) change their mechanical properties under the influence of the environment even in the absence of stress. For example, mechanical properties of soil, wood and some other materials depend on the level of humidity.

(g) *Damage of Concrete.* Since concrete is a non-homogeneous material there are zones of weak mechanical resistance in it, which under loading leads to the appearance of cracks.

1.2 On Damage Variables

The changes of the material structure mentioned above are, in general, irreversible; during the process of damage the entropy increases.

Damage accumulation can take place under elastic deformation (as in the case of high-cycle fatigue), under elastic-plastic deformation (as in the case of ductile plastic damage and low-cycle fatigue), under creep conditions (creep damage).

For simplicity, we assume that locally the temperature θ is constant, i.e., at the given point $\theta = const$.

For the phenomenological description of microstructural changes it is necessary to introduce, according to the principles of irreversible thermodynamics, some internal variables ("hidden parameters"), supplementing the set of thermodynamic ("observable") parameters of the basic undamaged state (elastic, plastic, creep, etc).

Let ε_{ij}, $i,j=1,2,3$ be the strain components, $\dot{\varepsilon}_{ij}$ strain rate components and σ_{ij} components of stress.

The set of parameters describing the damage is characterized by some mathematical entity which we denote by ω. The material is undamaged if $\omega=0$. Because there is no recovery ω is an increasing quantity.

The choice of the damage parameter, as a rule, is not simple. It can be done either by a physical microstructural analysis, or by a direct generalization of experimental data.

From the point of view of applications it is important that the set of functions ω should be simple enough and should have an evident mechanical sense. As a rule, the direct measuring of damage is impossible. But damage can be approximately estimated by observing the material properties (elastic, electric, acoustic, etc.).

We consider the damage variable as a surface density of intersections of cracks and cavities.

For different types of defects and their distribution it is expedient to use different mathematical descriptions of damage.

In the simplest case ω is a scalar function. In a more complex case ω is a vector function. The symmetric crack density tensor of second rank, introduced by Vakulenko and Kachanov [43] was used by Muzakami and Ohno [3] to describe the creep damage in metals; here, damage is characterized by six independent functions (components). Generally, damage affects the elastic properties of a material. As the damage develops, the material becomes anisotropic. Based on this fact, damage can be characterized (Cordebois and Sidoroff [5]) by a tensor of the fourth rank, that corresponds to the tensor of elastic coefficients and contains 21 independent components.

Note, in conclusion, that there is a more general mathematical representation of damage, proposed by Tamuzh and Lagsdinsh [6]; they describe damage by a set of functions given on the surface of a unit sphere surrounding the considered point of the body.

In this work we consider only the simple types of damage description.

For the complete analysis of the process of damage it is necessary to introduce the kinetic ("evolutional") equation in the form

$$\dot{\omega}=\dot{\omega}(\sigma_{ij},\omega,\ldots) \tag{1.1}$$

where

$$\dot{\omega}=\frac{d\omega}{d\lambda}$$

and where $\lambda>0$ is a monotomically increasing parameter similar to the time t.

In the case of an irreversible process such a parameter associated with the entropy s may be introduced in the form [7]

$$\lambda=\int_0^t s(\tau)d\tau \tag{1.2}$$

On the scale of this "time" any reversible process is instantaneous.

For the medium that is subjected to creep and for a visco-elastic medium it is expedient to consider the physical time t. In the case when plasticity does not depend on the strain rate, the parameter of loading can be considered as the parameter λ.

1.3 Isotropic Damage

Let us consider a certain section S of the body with a unit normal $\underset{\sim}{\nu}$ Figure 1.1. Let A_o be the initial area of the undamaged section. As a result of damage a certain part of the section is fractured ("lost"); let us denote the lost area by A.

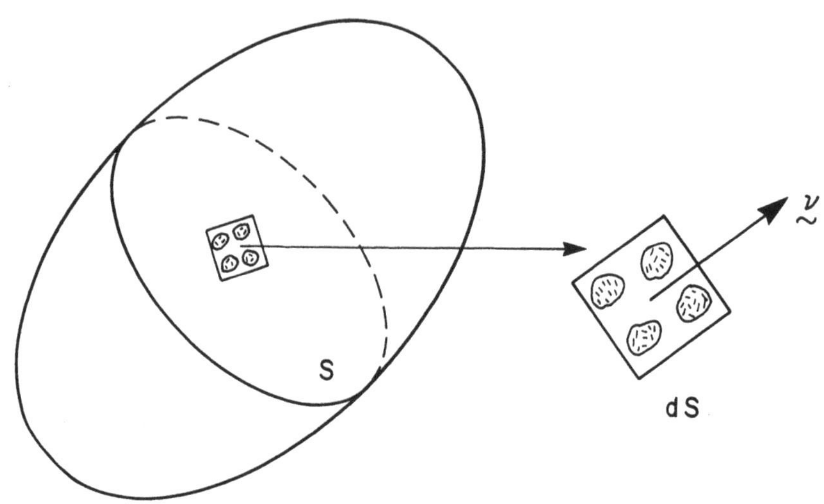

Figure 1.1

We are speaking here about local damage characteristics, but, of course, the values A_o, A and the normal ν are to be understood in the sense of appropriate averaging.

The value $A_o - A$ can be interpreted as the actual ("material") area of the section.

(a) *Isotropic Damage.* Now, let us consider isotropic damage. In this case cracks and voids are equally distributed in all directions. The damage variable can be considered as a scalar. We define the damage ω as

$$\omega = \frac{A}{A_o} \; ; \; 0 \leq \omega \leq 1 \tag{1.3}$$

It is a positive monotonically increasing function, i.e., $\dot{\omega} > 0$.

Sometimes it is convenient to use the function

$$\psi = 1 - \omega = \frac{A_o - A}{A_o} \tag{1.4}$$

This function is called "continuity". It is a positive monotonically decreasing function, i.e., $\dot{\psi}<0$; note that $1\geq\psi\geq0$.

For undamaged material, $\omega=0$ (or $\psi=1$); at fracture $\omega=1$ (or $\psi=0$).

(b) *Concept of Actual Stress.* Let us introduce the actual stress σ_a in the case of uniaxial tension

$$\sigma_a=\frac{P}{A_o-A}=\frac{P}{A_o(1-\omega)}=\frac{P}{A_o\psi}=\frac{\sigma}{\psi}$$

where σ is the nominal stress; σ_a is the stress related to the undamaged ("actual") area of the section, hence

$$\psi=\frac{\sigma}{\sigma_a}\,. \tag{1.5}$$

We assume that the strain response of the body is modified by damage only through the actual stress [31].

We also assume that the rate of damage growth is determined primarily by the level of the actual stress.

Thus, the stress-strain behavior of the damaged material can be represented by the constitutive equation of the virgin material (without damage) with the stress in it replaced by the actual stress.

These assumptions demonstrate the major role of the concept of the actual stress.

According to this concept the elastic strain of a damaged material is

$$\varepsilon=\frac{\sigma_a}{E}=\frac{1}{E}\frac{\sigma}{\psi}\,,\text{ or }\psi=\frac{1}{E}\frac{\sigma}{\varepsilon}\,. \tag{1.6}$$

So, the Hooke's law here has its usual form with the Young's modulus E being replaced by $E'=E\psi$.

In the case of an elastic-plastic deformation when the damage is a result of a large strain, it is natural to assume that damage does not depend on the elastic strain, hence

$$\frac{d\psi}{d\varepsilon}=0\,.$$

This condition leads to the relation

$$\frac{\sigma}{\varepsilon}=\frac{d\sigma}{d\varepsilon}\,.$$

Thus,

$$\psi = \frac{1}{E}\frac{d\sigma}{d\varepsilon} = \frac{E'}{E}\,. \tag{1.7}$$

Hence, the damage may be estimated by measuring the elastic response. Note that E' can be identified with the unloading modulus, Figure 1.2. An example of the experimental determination of damage for a light alloy [4] is shown in Figure 1.3.

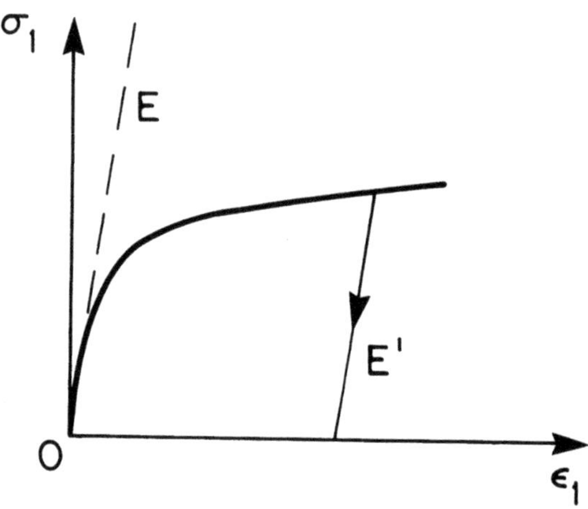

Figure 1.2

1.4 Kinetic Equation of Damage and Principle on Linear Summation

Let us consider a simple form of the kinetic equation in the case of uniaxial tension:

$$\frac{d\psi}{dt} = -A\left(\frac{\sigma}{\psi}\right)^n, \tag{1.8}$$

where $A>0$ and $n \geq 1$ are material constants. This equation, in general, agrees with the experimental data for metals. For the virgin material

$$\psi = 1 \text{ at } t = 0\,. \tag{1.9}$$

At the moment of complete deterioration ("fracture") $\psi=0$.

Let the rod be subjected to a constant tensile load. Then, in the case of small strain $\sigma = const. = \sigma_o$.

Integrating equation (1.8) with the initial condition (1.9) and taking $\psi=0$ we obtain the fracture time

$$t' = [A(n+1)\sigma_o^n]^{-1}\,. \tag{1.10}$$

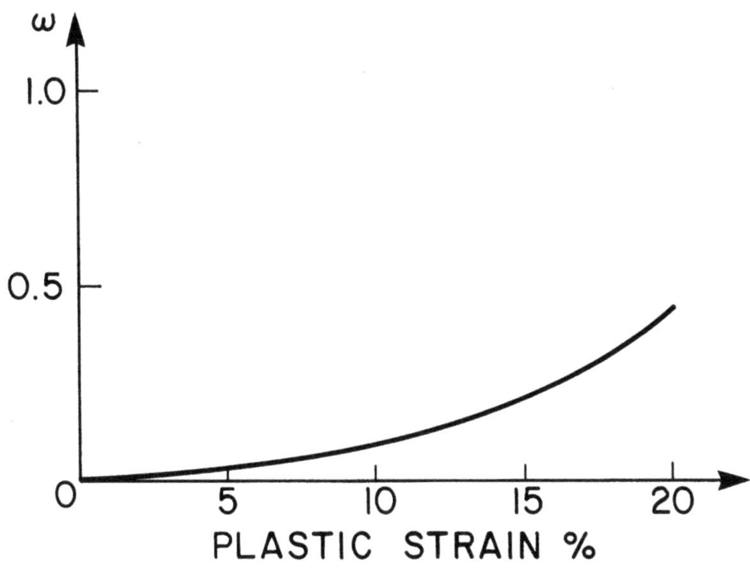

Figure 1.3

Consider the case of step-wise loading. At the time interval $\Delta t_k = t_k - t_{k-1}$, $k=1,2,...,s$ the stress is constant and equal to σ_k.

Integrating the kinetic equation for each step, using the condition of continuity of ψ, the condition (1.9) and the fracture condition, i.e.,

$$\psi = 0 \text{ at } t = t_s$$

we obtain

$$\sum_{k=1}^{s} \frac{\Delta t_k}{t'_k} = 1 \; ; \; t'_k = [A(n+1)\sigma_k^n]^{-1} \tag{1.11}$$

This relation is called the principle of linear summation of damage and is widely used in engineering.

If the loading is continuous, we consider infinitesimal steps dt. Then, instead of (1.11), we obtain

$$\int_o^{t'} \frac{d\tau}{t'(\tau)} = 1 \; ; \; t'(\tau) = [A(n+1)\sigma^n(\tau)]^{-1} \tag{1.12}$$

Thus, the principle of linear summation is a consequence of the kinetic equation.

Odquist and Hult [11] emphasized the equivalence of this principle and the kinetic equation (1.8).

Indeed, in the case of a power law, the kinetic equation may be written in the form

$$\frac{d\bar{\psi}}{dt}=-A(n+1)\sigma^n \ , \tag{1.13}$$

where $\bar{\psi}=\psi^{n+1}$. If $t'(t)$ is the time to fracture under the action of the stress $\sigma(t)$, equation (1.13) takes the form

$$\frac{d\bar{\psi}}{dt}=-\frac{1}{t'(t)} \tag{1.14}$$

and has the solution

$$\bar{\psi}=1-\int_o^t\frac{d\tau}{t'(\tau)} \ . \tag{1.15}$$

The level of damage can be charactrized by the function $\bar{\psi}$. If $t<t'$ the relation (1.12) leads to the equation (1.15). Differentiating the latter, we obtain equation (1.14).

1.5 Elastic Body with Damage

Consider a linear elastic isotropic body. Under loading there is a nucleation and growth of damage due to non-homogeneity of the material and initial defects. Let the damage be isotropic. The thermodynamic parameters of the state are the strain components ε_{ij} and the continuity ψ. In the case of constant temperature the density of Helmholtz's free energy is

$$f=f(\varepsilon_{ij},\psi) \ .$$

Hence,

$$df=\frac{\partial f}{\partial\varepsilon_{ij}}d\varepsilon_{ij}+\frac{\partial f}{\partial\psi}d\psi \ .$$

The derivatives $\partial f/\partial\varepsilon_{ij}$ are equal to the corresponding generalized forces, i.e. the stress components σ_{ij}. We denote the derivative $\partial f/\partial\psi$ by $-Q$; Q is the generalized force associated with the damage parameter ψ. Thus,

$$df=\sigma_{ij}d\varepsilon_{ij}-Qd\psi \ . \tag{1.16}$$

For the elastic body without damage $df=\sigma_{ij}d\varepsilon_{ij}=dU$, where $U(\varepsilon_{ij})$ is the density of the elastic strain energy.

In the damaged state, according to the concept of actual stress, the stress components are related to the strain components by Hooke's law with the Young's modulus E being replaced by $E'=E\psi$. Hence, $\sigma_{ij}d\varepsilon_{ij}=\psi dU$.

It is not difficult to see that in this case

$$f = U\psi \tag{1.17}$$

and

$$Q = -U\,. \tag{1.18}$$

The second term in equation (1.16) characterizes the increment of entropy due to the increment of damage. The second law of thermodynamics requires that

$$Q d\psi > 0\,.$$

Since $d\psi < 0$ we get $Q < 0$.

Analogously to the J-integral in the fracture mechanics the thermodynamic force Q can be interpreted [31] as the energy release rate resulting from the growth of damage.

1.6 Damage in Creep Conditions

In creep conditions the entropy production rate density is

$$\dot{s} = \sigma_{ij}\dot{\varepsilon}_{ij} + Q_* \dot{\psi} \tag{1.19}$$

where Q_* is the thermodynamic force associated with the rate $\dot{\psi}$.

The first term in (1.19) represents the dissipation due to creep as a non-linear viscous flow; the second term characterizes the dissipation due to the process of damage.

Note that in the case of steady creep (see, for example, [9]),

$$s_{ij} = \frac{\partial L}{\partial \dot{\varepsilon}_{ij}} \tag{1.20}$$

where s_{ij} are the components of the stress deviator, and $L(\dot{\varepsilon}_{ij})$ is a certain potential. In the case of the power creep law, L is a homogeneous positively definite function of degree $\mu + 1$, where the index μ is in the interval $0 < \mu \leq 1$.

In the case of incompressibility,

$$\sigma_{ij}\dot{\varepsilon}_{ij} = s_{ij}\dot{\varepsilon}_{ij}\,.$$

Consequently, according to the Euler's theorem,

$$\sigma_{ij}\dot{\varepsilon}_{ij} = (\mu + 1)L\,. \tag{1.21}$$

Thus, the first term in (1.19) is positive. The second law of thermodynamics requires that $\dot{s} > 0$. Hence,

$$Q_* < 0 \,. \tag{1.22}$$

The thermodynamic force Q_* characterizes the dissipation rate due to the rate of damage.

Chapter 2

CREEP AND FRACTURE UNDER UNIAXIAL STRESS

2.1 Creep Under Uniaxial Tension

(a) *Creep under a Constant Load.* At constant load and temperature the typical creep curve with primary portion AB, secondary portion BC and tertiary portion CD has the form shown in Figure 2.1. In the secondary portion the strain rate is approximately constant ("steady creep"). At the moment of loading the specimen obtains an instantaneous elastic or elastic-plastic strain ε^o. The creep acceleration in the third portion depends on the brand of metal and on the temperature.

In the case of ductile failure the tertiary portion features the decrease of the cross-section and the formation of a neck. In the well-known Andrade's creep tests at constant stress, the third portion was not observed.

In the case of brittle fracture, the third portion is characterized by the development of damage in the metal.

At constant temperature, the form of the creep curve depends on the stress level (see Krish [13], for example).

At high stress ((short-time creep), Figure 2.2 a), fracture is usually ductile. It occurs at a large strain, the creep acceleration in the third portion being intensive.

At the low stress ((long-time creep), Figure 2.2 b), fracture is brittle. It occurs at a small strain, with the creep acceleration in the third portion not so intensive, for some metals (e.g. for nickel base alloy) in fact insignificant.

The experimental data characterizing the relationship between the stress σ and the time t_* to fracture are usually entered on a log-log plot Figure 2.3.

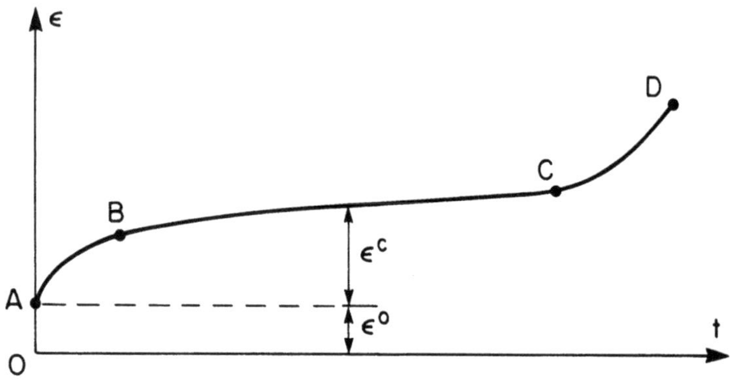

Figure 2.1

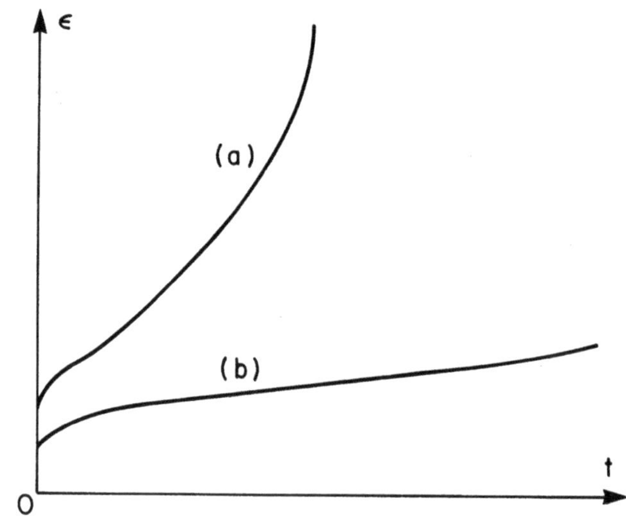

Figure 2.2

This creep rupture curve has, as a rule, a "point" of change of slope; the left-hand side of the curve represents ductile failure, the right-hand side - brittle fracture.

(b) *Steady Creep Theory*. Neglecting the short primary portion, we obtain the simplest equation of creep:

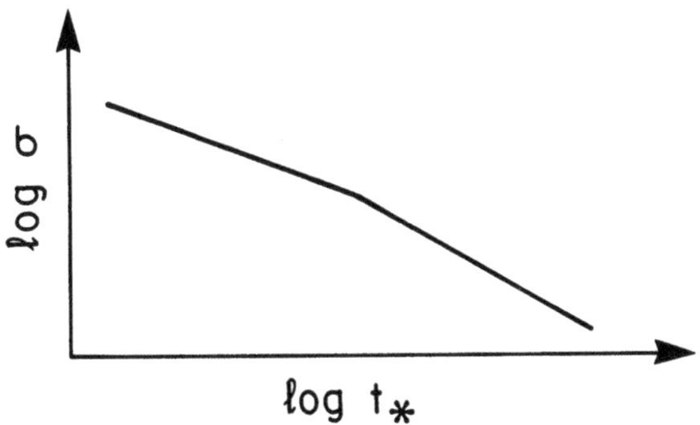

Figure 2.3

$$\dot{\varepsilon}^c = f(\sigma) \,. \tag{2.1}$$

In the case of $f(\sigma)=k\sigma$, where k=constant, we have a linearly viscous medium (Newtonian fluid). In the case of creep of metals, the creep strain rate is a non-linear function of stress, Figure 2.4. The power approximation (power law, Norton's law)

$$\dot{\varepsilon}^c = B_1\sigma^m \,, \tag{2.2}$$

where B_1,m are materials constants, is often used.

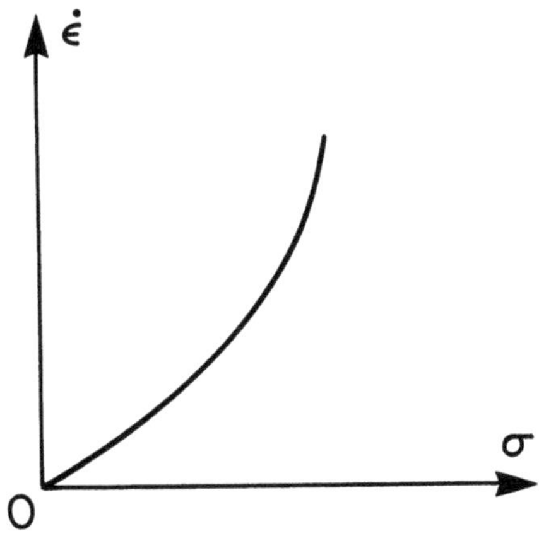

Figure 2.4

(c) *Hardening Theory*. Hardening is analytically given by

$$\dot{\varepsilon}^c = f(\sigma, \varepsilon^c) . \tag{2.3}$$

Note, that this relationship can be of the type

$$\dot{\varepsilon}^c = \frac{f_1(\sigma)}{f_2(\varepsilon^c)} , \tag{2.4}$$

where $f_1(\sigma) \geq 0$ and $f_2(\varepsilon^c) > 0$ are monotonically increasing functions.

(d) *Creep-Plastic Body*. At high stress, creep is accompanied by instantaneous plastic strain ε^p. The total strain rate is

$$\dot{\varepsilon} = \dot{\varepsilon}^c + \dot{\varepsilon}^p . \tag{2.5}$$

The creep strain rate $\dot{\varepsilon}^c$ is described by the flow theory (2.1) or by the hardening theory (2.3). The plastic strain rate is

$$\dot{\varepsilon}^p = \phi(\sigma) \frac{d\sigma}{dt} , \tag{2.6}$$

where $\phi(\sigma)$ is a non-negative monotonically increasing function.

If the plastic strain hardening is small, it is expedient to use the scheme of ideally-plastic behavior. Then

$$\sigma = const. = \sigma_y , \tag{2.7}$$

where σ_y is the yield stress.

(e) *Creep-Elastic Body*. In the case of low stress and small strain, it is necessary in some cases to take the elastic strain ε^e into consideration. Then, the total strain rate is

$$\dot{\varepsilon} = \dot{\varepsilon}^c + \dot{\varepsilon}^e . \tag{2.8}$$

According to Hooke's law,

$$\dot{\varepsilon}^e = \frac{1}{E} \frac{d\sigma}{dt}$$

where E is Young's modulus. In the case $m=1$ we have a linear visco-elastic body (Maxwell's medium).

2.2 Time to Ductile ("Viscous") Fracture

(a) *Hoff's Solution*. Consider the problem of intensive creep of a rod under the action of a constant tensile P. Denote the initial length and area of the cross-section of the rod by ℓ_o, F_o, and the current length and area by ℓ, F. Then $\sigma = P/F$ is the current stress and $\sigma_o = P/F_o$ is the initial stress. The strain rate and the strain are

$$\dot{\varepsilon}=\frac{1}{\ell}\frac{d\ell}{dt}\ ;\ \varepsilon=\ell n\frac{\ell}{\ell_o}\ .$$

According to the flow theory,

$$\frac{1}{\ell}\frac{d\ell}{dt}=f(\sigma)\ . \tag{2.9}$$

Let us introduce $\ell/\ell_o=\lambda$. The condition of incompressibility is

$$F\ell=F_o\ell_o\ .$$

Therefore,

$$\sigma=\frac{P}{F}=\sigma_o\lambda\ .$$

Thus, we obtain the differential equation

$$\frac{1}{\lambda}\frac{d\lambda}{dt}=f(\sigma_o\lambda) \tag{2.10}$$

and the initial condition at $t=0$

$$\lambda=1 \text{ at } t=0\ . \tag{2.11}$$

Let us introduce the function

$$\Phi(z)=\int_o^z\frac{dx}{xf(x)}\ .$$

Then the solution of the differential equation has the form

$$t=\Phi(\sigma_o\lambda)-\Phi(\sigma_o)\ .$$

As $\lambda\to\infty$ the function $f(\lambda\sigma_o)$ increases not slower than λ^k, where $k>0$ is a certain number. The integral is convergent; let its limit value be denoted by $\Phi(\infty)$.

Hence, there is a finite time $t=t_1$ at which the rod stretches into an infinitely thin thread. If follows from (2.10) that as $t\to t_1$ we have $d\lambda/dt\to\infty$. Thus

$$t_1=\Phi(\infty)-\Phi(\sigma_o)\ . \tag{2.12}$$

Hoff considered the time t_1 as the time to fracture, because at this time the rod loses its resistivity.

Note that the process of uniform extension is usually arrested with the formation of a neck.

In the special case of the power law we have $f(x)=B_1x^m$ and

$$t_1=\frac{1}{m\dot{\varepsilon}_o}(1-\lambda^{-m}) \tag{2.13}$$

where $\dot{\varepsilon}_o=B_1\sigma_o^m$ is the initial creep strain rate.

Taking $\lambda\to\infty$ we obtain the time to viscous fracture:

$$t_1=\frac{1}{m\dot{\varepsilon}_o}\ . \tag{2.14}$$

From (2.13) and the condition of incompressibility it follows that

$$\frac{F}{F_o}=(1-\frac{t}{t_1})^{1/m}\ . \tag{2.15}$$

For $m>1$

$$\frac{d}{dt}(\frac{F}{F_o})\to\infty \text{ as } t\to t_1\ .$$

The relation (2.15) is shown in Figure 2.5

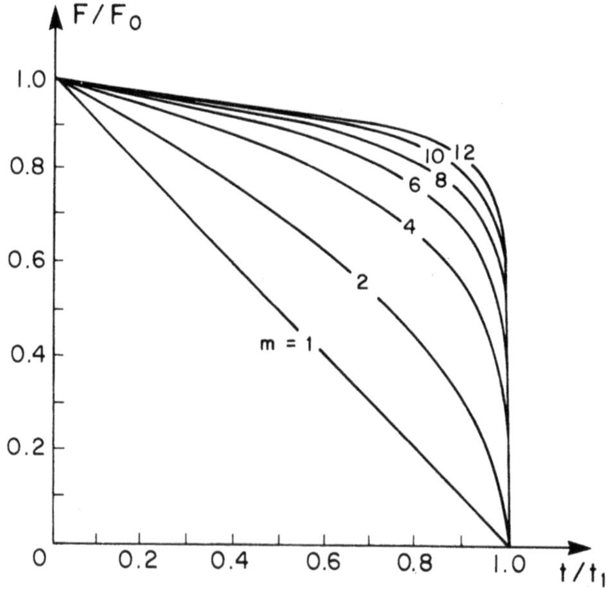

Figure 2.5

If $m \gg 1$, the cross-section decreases fast only in the last period of the "life" of the sample. The relation (2.13) can be written in the form

$$\frac{t}{t_1} = 1 - \lambda^{-m} . \tag{2.16}$$

In general, the simple result (2.14) agrees with the experimental data on creep ductile failure.

If $m \gg 1$, the right-hand side of (2.16) is close to unity even for a relatively small elongation. Therefore, practically, the fracture condition $\lambda \to \infty$ may be replaced by the condition $\lambda \to \lambda_* < \infty$ where the limit elongation λ_* can be considered as a material constant.

Now, consider the problem of extension of the rod proceeding from the hardening theory of creep. In this case,

$$\frac{1}{\lambda}\frac{d\lambda}{dt} = \frac{f_1(\sigma_o \lambda)}{f_2(\ell n \lambda)} \equiv \frac{f_1(\sigma_o \lambda)}{f_2^*(\lambda)} . \tag{2.17}$$

Hence, under the same initial condition, we obtain

$$t = \int_1^\lambda \frac{f_2^*(x)dx}{x f_1(\sigma_o x)} .$$

Introducing the function

$$\Psi(z) = \int_1^z \frac{f_2^*(x)dx}{x f_1(\sigma_o x)}$$

we can rewrite the previous relation in the form

$$t = \Psi(\lambda) - \Psi(1) .$$

The time to ductile fracture is

$$t_1 = \Psi(\infty) - \Psi(1) . \tag{2.18}$$

(b) *Time to Fracture for a Creep-Plastic Rod.* Let us consider the behavior of the rod using the equations (2.5) and (2.6). Then

$$\dot{\varepsilon} = f(\sigma) + \phi(\sigma)\frac{d\sigma}{dt} . \tag{2.19}$$

Introducing the notations

$$\lambda = \frac{\ell}{\ell_o} ; \dot{\varepsilon} = \frac{1}{\lambda}\frac{d\lambda}{dt} ; \sigma = \frac{P}{F_o}\lambda ,$$

and considering the case of a constant load, $P/F_o = const = \sigma_o$, we obtain

$$[1-\phi(\sigma_o\lambda)\sigma_o\lambda]\frac{d\lambda}{dt}=f(\sigma_o\lambda)\lambda .$$

Since $\phi\geq 0$, $f\geq 0$, $d\lambda/dt>0$, we have

$$[1-\phi(\sigma_o\lambda)\sigma_o\lambda]\geq 0 .$$

The condition

$$1-\phi(\sigma_o\lambda)\sigma_o\lambda=0$$

determines the unique root λ^*, Figure 2.6.

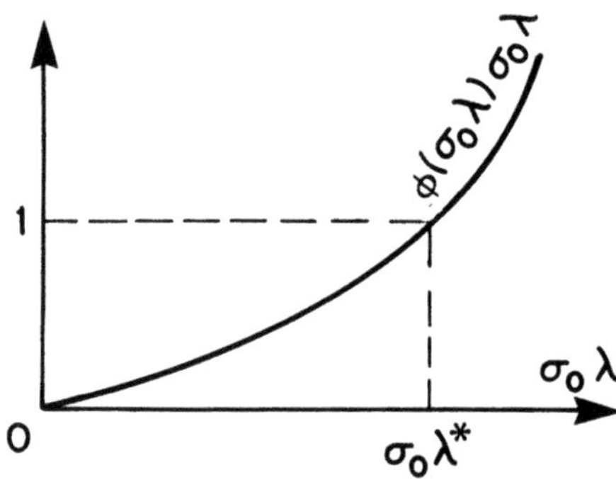

Figure 2.6

As $\lambda\to\lambda^*$, the elongation rate $\dot{\lambda}\to\infty$. The corresponding time t_1 can be considered as the time to ductile fracture.

If $\phi\neq 0$, the length of the rod cannot increase indefinitely. At the moment of loading, the instantaneous elongation λ_o takes place. Then, the rod extends, primarily because of creep. But, simultaneously, some plastic elongation takes place caused by both the decrease of the cross-section and the corresponding increase of stress. At the time moment t_1, the critical length λ^* is reached and the uniform extension becomes unstable.

Finally, consider the case of *ideal plasticity.*

If the stress $\sigma<\sigma_y$ only creep takes place, and the stress is equal to $\sigma=\sigma_o\lambda<\sigma_y$.

The plastic failure is reached for

$$\lambda_y=\frac{\sigma_y}{\sigma_o}>1 .$$

The corresponding time to fracture is

$$\bar{t}_1 = \Phi(\sigma_y) - \Phi(\sigma_o) . \tag{2.20}$$

If $\sigma_o = \sigma_y$, the rod is fractured instantaneoulsy.

In the case of the power law,

$$\bar{t}_1 = t_1[1 - (\frac{\sigma_o}{\sigma_y})^m], \tag{2.21}$$

where $\bar{t}_1$ is defined by (2.14). For large values of the index m the correction is small. This result can be explained by the fact that the stress increases fast in the final stage of the creep flow, Figure 2.5.

Formula (2.21) was derived by V. Rosenblum. Note, that for viscous fracture in the case of step-wise loading, the principle of linear summation is also true.

(c) *Example. Ductile Fracture of a Framework, (Figure 2.7).* The initial dimensions of the bars are denoted by ℓ_o, F_o, the current values by ℓ, F. The condition of incompressibility is $\ell F = \ell_o F_o$. The stress in the bars is

$$s = \sigma_o [1 - (\frac{a}{\ell})^2]^{-1/2} ;$$

$$s_o = \sigma_o [1 - (\frac{a}{\ell_o})^2]^{-1/2} .$$

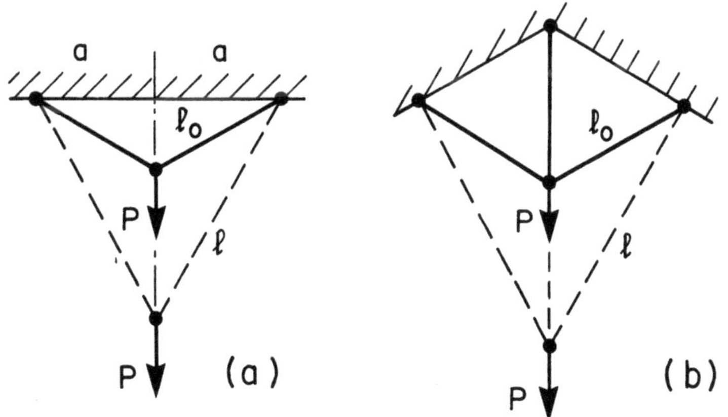

Figure 2.7

The creep equation is

$$\frac{1}{\ell}\frac{d\ell}{dt} = B_1 \sigma_o^m \{\frac{1}{\ell_o}[1 - (\frac{a}{\ell})^2]^{-1/2}\}^m ,$$

where $\sigma_o = P/2F_o$ indicates the stress as $a=0$.

Integrating and substituting

$$\frac{\ell}{a} = \lambda \, ; \; \frac{\ell_o}{a} = \lambda_o \, ;$$

$$J(\lambda) = \int_{\lambda_o}^{\lambda} (\lambda^2 - 1)^{m/2} \lambda^{-2m-1} d\lambda \, ,$$

we obtain

$$J(\lambda) = \frac{\lambda_o^{-m}}{m} \frac{t}{t_o} \, ,$$

where $t_o = [mB_1\sigma_o^m]^{-1}$ is the time to ductile fracture of the generating framework as $a=0$. Consider the special case $m=4$. Then,

$$\left(\frac{\lambda_o}{\lambda}\right)^4 \left(-1 + \frac{4}{3\lambda^2} - \frac{1}{2\lambda^4}\right) + \left(1 - \frac{4}{3\lambda_o^2} + \frac{1}{2\lambda_o^4}\right) = \frac{t}{t_o} \, . \tag{2.22}$$

Fracture occurs when $\lambda \to \infty$. Hence, the time to ductile fracture is

$$t_{11} = t_o \left(1 - \frac{4}{3\lambda_o^2} + \frac{1}{2\lambda_o^4}\right) . \tag{2.23}$$

This dependence is shown in Figure 2.8. Taking into consideration that the stress is bounded by the value of yield stress σ_y, we find that the maximum value λ_* of the length parameter is defined by the relation

$$\lambda_o \frac{\sigma_y}{\sigma_o} = \lambda^2 (\lambda^2 - 1)^{-1/2} \tag{2.24}$$

Substituting the value λ_* into equation (2.23), we find the corresponding time to fracture; note that $\lambda \geq \lambda_o > 1$.

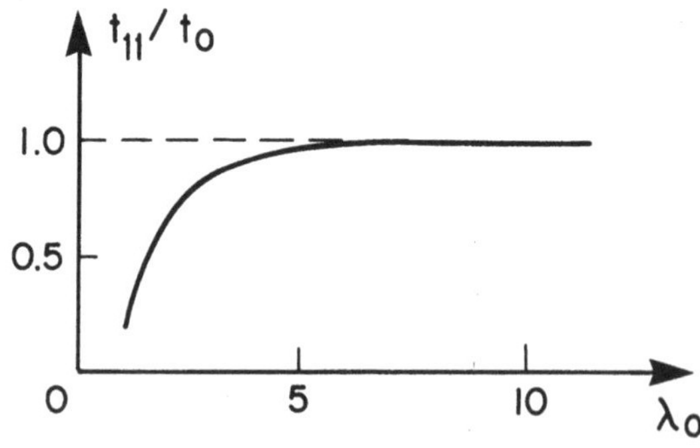

Figure 2.8

The relationship (2.24) is shown in Figure 2.9 by a solid line. If $\lambda_o\sigma_y/\sigma_o<2$, equilibrium is impossible, there is an accelerated plastic flow; if $\lambda_o\sigma_y/\sigma_o>2$, while $\lambda_o>\sqrt{2}$, the framework undergoes creep when $\lambda<\lambda_*$. At $\lambda=\lambda_*$ plastic failure occurs. If $\lambda_o\sigma_y/\sigma_o>2$, but $\lambda_o<\sqrt{2}$, the framework undergoes under sufficiently large initial stress initial plastic deformation, after which creep develop as $\lambda<\lambda_*$.

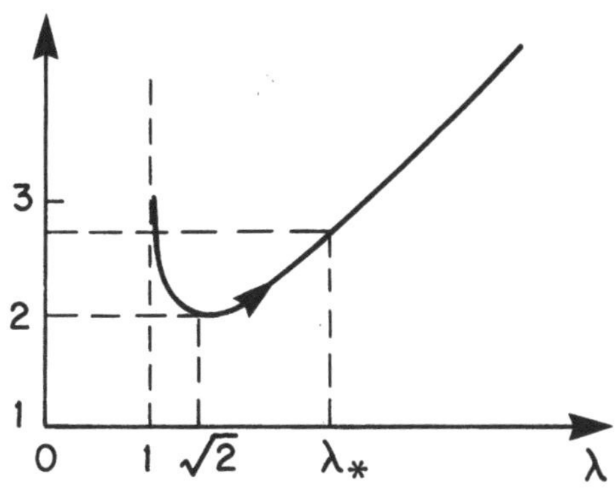

Figure 2.9

2.3 Time to Brittle Fracture

(a) *Brittle Fracture under Constant Load.* Consider a cylinder bar under a constant tensile load P. We assume that the stress σ is considerably less than the ultimate strength. According to the kinetic equation (1.8), we set

$$\frac{d\psi}{dt}=-A\left(\frac{\sigma}{\psi}\right)^n . \tag{2.25}$$

In the case of brittle fracture the strain is small, hence, $\sigma=\sigma_o$. Integration under the initial condition leads to

$$1-\psi^{n+1}=A(n+1)\sigma_o^n t \tag{2.26}$$

At the moment t' of fracture the continuity $\psi=0$; hence,

$$t'=[(n+1)A\sigma_o^n]^{-1} \tag{2.27}$$

and

$$t=t'(1-\psi^{n+1})\ ;\ \psi=\left(1-\frac{t}{t'}\right)^{1/n+1} . \tag{2.28}$$

It is possible to assume that at a certain level $\psi=\psi_o$, the damage is localized and macrocracks occur; the corresponding time according to the solution (2.26) is

$$t'_o = t'(1-\psi_o^{n+1}) . \tag{2.29}$$

Since $0<\psi_o<1$ and, generally, $n>>1$, the correction is insignificant. Sometimes, it is expedient to evaluate ψ_o from experimental data. Figure 2.10 shows the dependence of the fracture time t' on the stress σ_o; σ_u is the ultimate strength. The dashed line characterizes the actual dependence close to the point σ_u.

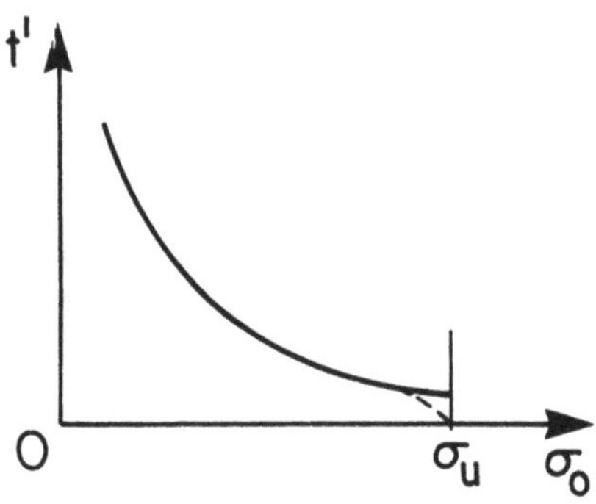

Figure 2.10

(b) *Example.* As an example consider the case of brittle fracture of a system consisting of three bars of equal length ℓ and equal cross-sectional area F, Figure 2.11. In the elastic state the stress in the vertical bar is $\sigma_1 = 2/3s$, where $s = P/F$; the stress in the lateral bars is $\sigma_2 = 1/3s$. The vertical bar will be fractured at the moment $t'_{11} = [A(n+1)\sigma_1^n]^{-1}$. Then only the lateral bars will support the load.

The stress in these bars will increase to the value s. According to the principle of linear summation the fracture time t' of the system is given by the relation

$$\frac{t'_{11}}{t'_1} + \frac{t'_1 - t'_{11}}{t'_2} = 1$$

where t'_1 is the fracture time of the lateral bar under the stress σ_2 in the time interval $0 \leq t \leq t'_{11}$; t'_2 is the fracture time of the lateral bar under the stress s (for $t > t'_{11}$). Note that we can consider the stress state of the system as elastic only if the creep strain is very small. In the case of developed creep, the stress state is close to the steady creep stress state.

(c) *Brittle Fracture in Relaxation.* In the case of relaxation, the total strain $\varepsilon = const.$; hence $\dot{\varepsilon}^e + \dot{\varepsilon}^c = 0$, or

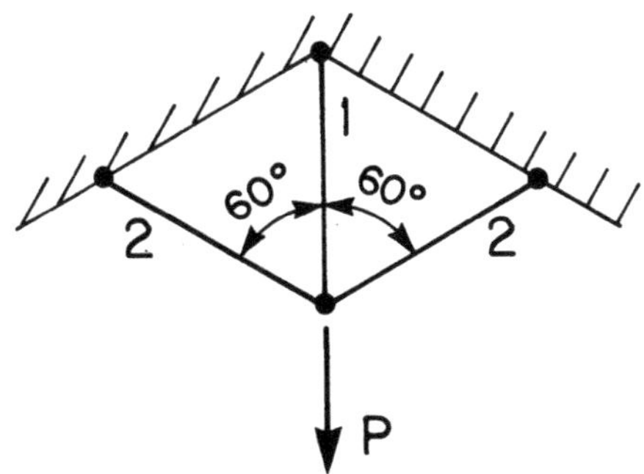

Figure 2.11

$$\frac{1}{E}\frac{d\sigma}{dt}+B_1\sigma^m=0\,. \tag{2.30}$$

Integrating the equation under the initial condition $\sigma=\sigma_o$ at $t=0$ we obtain the relaxation curve in the form

$$\sigma(t)=\sigma_o\rho(t)$$

where $\rho(t)$ is a monotonically decreasing function; $\rho(0)=1$.

Substituting $\sigma(t)$ into the kinetic equation, integrating, and assuming that at the fracture moment t'_r the continuity $\psi=0$, we find that

$$t'_r=t'\int_o^{t'_r}\rho^n(\tau)d\tau \tag{2.31}$$

where t' is the fracture time under constant stress σ_o.

If the process of relaxation is a slow one, there is a finite fracture time t'_r. For a fast decreasing stress, the root t'_r of the above written equation does not always exist.

It is also possible to consider the case when during the relaxation process the stress increases at certain moments of time (as in bolt-tightening, for example).

(d) *Brittle Fracture under a Constant Strain Rate.* In the case of a sample being subjected to a constant strain rate (with the velocity v given), we have

$$\frac{d\ell}{dt}=const.=v\,.$$

Then the length ℓ of the sample is

$$\ell = \ell_o + vt \ .$$

Making use of the steady creep equation

$$\dot{\varepsilon} = B_1 \sigma^m \ ,$$

we find the stress σ as a function of the strain rate $\dot{\varepsilon} = v/\ell$. Substituting σ into the damage equation (2.25) and using the incompressibility condition, we find that the moment of fracture $t = t_*$ (at $\psi = 0$) is given by the relation

$$t' = \int_o^{t_*} (1 - \frac{v}{\ell_o})^{-n/m} d\tau$$

where t' is the time to brittle fracture under the constant stress σ_o; σ_o is the initial stress (when $\ell = \ell_o$, $\dot{\varepsilon} = v/\ell_o$).

In the case $n < m$ we obtain

$$t_* = \frac{\ell_o}{v} [(1 + \kappa \alpha \sigma_o^{m-n})^{n/m-n} - 1] \ , \tag{2.32}$$

where

$$\alpha = \frac{m-n}{m}$$

and

$$\kappa = \frac{B_1}{A(n+1)} \ .$$

At the moment of fracture, the strain is equal to

$$\varepsilon_* = \frac{v}{\ell_o} t_* \ .$$

The second term in the parenthesis is positive and $<<1$. Expanding the parenthesis into a series and retaining only the first two terms, we obtain

$$\varepsilon_* \approx \kappa (\frac{\dot{\varepsilon}_o}{B_1})^\alpha \ . \tag{2.33}$$

This result agrees with the experimental data of Staniukovich [14].

(e) *Brittle Fracture under Cyclic Loading.* The problem of brittle fracture in creep conditions under variable (mainly cyclic) loading is of great importance. It is well-known that it is difficult to describe in terms of mechanics. Consider the scheme applicable in the case of slow cyclic loading (low-cycle fatigue). Then, only the damage due to creep can be taken into account.

Note that the superposition of a hydrostatic pressure practically does not affect the rate of creep, but can slow down the process of damage; in bending tests, the damage does not develop in the compression zone. The "healing" of damage in compression of the previously extended metal is insignificant and can be neglected.

Thus, we obtain the scheme:

$$\frac{d\psi}{dt}=\begin{cases}-A\left(\frac{\sigma}{\psi}\right)^n & \text{for } \sigma>0,\\ 0 & \text{for } \sigma\leq 0.\end{cases} \tag{2.34}$$

It implies that all periods of compressive stress are totally inactive in promoting creep fracture. This scheme of behavior was termed by Hult and Broberg [15] "dormant damage mechanism".

(f) *Extension-Compression of a Rod.* Consider the problem of a rod subjected to extension-compression. Let the stress σ be changed harmonically

$$\sigma=\sigma_o(\alpha+\beta\sin 2\pi\frac{t}{T}) \tag{2.35}$$

where $\sigma_o>0$, $\alpha>0$, β and T are constants. If the values of constants are such that $\sigma>0$ the time to fracture t' is determined by the relation

$$1=(n+1)A\sigma_o^n\int_o^{t'}(\alpha+\beta\sin 2\pi\frac{t}{T})^n dt\ . \tag{2.36}$$

If $\alpha>>|\beta|$, the fracture time is close to the fracture time under constant stess $\sigma_o\alpha$. If $\sigma<0$ during the process, fracture does not occur.

Consider a general case when the stress changes the sign, (Figure 2.12 a). Introducing the Heaviside unit step function $H(x)$ that is equal to 1 for $x>0$ and to zero for $x<0$, we obtain the relation that determines the fracture time t',

$$t'_o=\int_o^{t'}|\alpha+\beta\sin 2\pi\frac{t}{T}|^n H(\sigma)dt\ , \tag{2.37}$$

where t'_o is the time to fracture under constant stress σ_o. Denoting

$$\frac{1}{T}\int_o^T|\alpha+\beta\sin 2\pi\frac{t}{T}|^n H(\sigma)dt=K\ ,$$

and assuming that the time to fracture contains a sufficiently great number of cycles, i.e., $t'\sim\kappa T$, where $\kappa=E(t'/T)>>1$ is the entire part of t'/T, we find that the number of cycles to fracture is

$$\kappa = E\left(\frac{t'_o}{KT}\right). \tag{2.38}$$

If $\alpha=0$, $\beta=1$ the cycles are symmetric; then

$$K = \frac{1}{2\pi}\int_o^T |\sin x|^n dx \; .$$

In the case of step-wise loading, (Figure 2.12 b) we have $K = T_1/T$; and the number of cycles to fracture is

$$\kappa = E\left(\frac{t'_o}{T_1}\right).$$

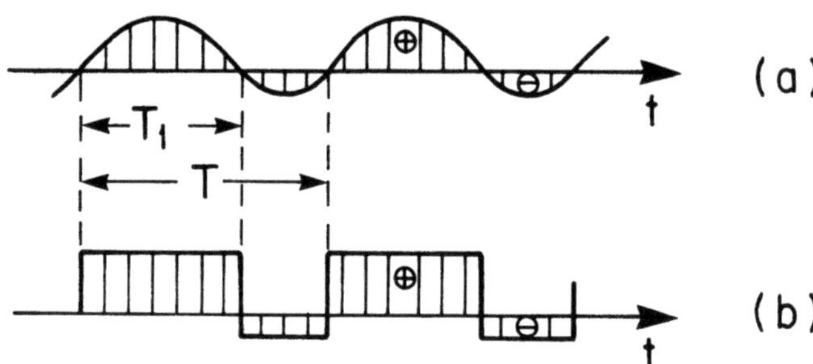

Figure 2.12

(g) *Influence of the Temperature on Brittle Fracture.* The time to brittle fracture is considerably affected by the temperature θ. Therefore, the problem of fracture of non-uniformly heated bodies becomes important. If the range of the temperatures is not too large, the following approximate scheme can be used. Let the kinetic equation be taken in its simplest form; then, the index n can be considered as constant within the given interval of temperatures and the coefficient A - as a function of temperature

$$A = A_o e^{\alpha\theta}, \tag{2.38}$$

where A_o and α are constants.

2.4 Brittle-Viscous Fracture

(a) *On Brittle-Viscous Fracture.* In section 2.2, the ductile ("viscous") fracture under creep conditions was considered. Such fracture is characterized by a high stress and a relatively short life-time; the dominant mechanism in this case is a slip within the grains of the metal. Under low stress, the dominant mechanism is the nucleation and growth of microcracks at the grain boundaries; the fracture is brittle and occurs at low strain.

Under an intermediate stress, the fracture is of mixed type. This is illustrated by the curve of long-time strength shown in Figure 2.3; the left-hand side of the curve corresponds to ductile fracture, the right-hand side to brittle fracture (sect. 2.3).

For the description of the entire curve, the process of fracture will be considered as growth of damage against the background of growing creep strain.

(b) *Extension of a Rod under a Constant Load.* We assume that the creep strain is not influenced by the process of damage. If the final creep stage characterized by significant dilatancy is excluded from the analysis, this assumption can be supported by the following considerations: firstly, the mechanisms of cracking and creep flow are, generally, different; secondly, the experimental creep curves give a summarized information. It is impossible to isolate from these curves the inherent creep.

We denote a current length of the rod and its cross-sectional area by ℓ and F, correspondingly; initially (at $t=0$) $\ell=\ell_o$, $F=F_o$. From the condition of incompressibility it follows that

$$\ell F=\ell_o F_o \ .$$

According to the creep law, the strain rate $\dot{\varepsilon}=1/\ell\, d\ell/dt$ is related to stress $\sigma=P/F$ by the equation

$$\frac{1}{\ell}\frac{d\ell}{dt}=B_1\sigma^m \ , \tag{2.39}$$

where B_1,m are constants (at the given temperature). From these equations we can find the time of viscous fracture t_1 that corresponds to the moment when F becomes zero (sec. 2.2):

$$t_1=\frac{1}{m\dot{\varepsilon}_o} \ ; \ \dot{\varepsilon}_o=B_1\sigma_o^m \ . \tag{2.40}$$

According to this solution

$$\frac{F}{F_o}=(1-\frac{t}{t_1})^{1/m} \ , \ t\leq t_1 \ . \tag{2.41}$$

Typically, $m>>1$, therefore a sharp decrease of the cross-sectional area occurs only during the last period of deformation (see Figure 2.5).

When there is no creep, $F=F_o$ and, integrating the kinetic equation of damage with the initial condition $\psi=1$, we obtain the time to brittle fracture as

$$t'=[(n+1)A\sigma_o^n]^{-1} \tag{2.42}$$

while the continuity ψ becomes zero. According to experimental data, the slope of the curve in logarithmic coordinates $\log\sigma_o$, $\log t_*$ stays constant or increases when the time to fracture t_* increases too; therefore, $m \geq n$.

In the general case, substituting the area F into the kinetic equation of damage,

$$\frac{d\psi}{dt} = -A\left(\frac{\sigma}{\psi}\right)^n , \tag{2.43}$$

we obtain a differential equation with separable variables. Integrating and assuming that $\psi = 0$, we find the fracture time t_* as

$$\frac{t_*}{t_1} = 1 - \left(1 - \frac{m-n}{m}\frac{t'}{t_1}\right)^{m/m-n} . \tag{2.44}$$

This solution makes sense if $t_* \leq t_1$ (otherwise the brittle fracture does not occur). Therefore,

$$\sigma_o \leq \left(\frac{1}{\kappa(m-n)}\right)^{1/m-n} \equiv \bar{\sigma}_o \; ; \; \kappa = \frac{B_1}{A(n+1)} . \tag{2.45}$$

Under a higher stress there occurs viscous fracture (according to Hoff's solution); at a lower stress, fracture is brittle, but it takes place not at a low level of strain; the latter can be found from (2.41) at $t = t_*$. The curve of long-time stength corresponding to the obtained solution, is shown in Figure 2.13 by the solid line. The straight lines abc and de characterize the viscous and brittle fractures according to equations (2.14) and (2.27). For the stress $\sigma_o > \bar{\sigma}_o$ the curve of long-time strength coincides with the line ab; at a lower stress, the brittle mode of fracture becomes more important and the curve approaches the line de. As has been mentioned, the constants A and n can be determined from long-time strength experiments (in the case of small strain).

In the case of $m = n$ the solution takes the form

$$\frac{t_*}{t_1} = q_1 \; ; \; q_1 = 1 - exp(-m\kappa) \leq 1 . \tag{2.46}$$

In this case, fracture is brittle (i.e., it is determined by the condition $\psi = 0$) and occurs at one and the same strain

$$q_1 = 1 - \left(\frac{\ell_o}{\ell_*}\right)^m$$

where ℓ_* is the length of the rod at the moment of fracture.

If $m = n$, the metal is stable; the experimental curve does not sharply change its direction. The elongation of the rod at the moment of fracture can be well described by the choice of the constant A.

The solution given above provides a good qualitative description of the experimental data; quantitatively, there can be deviations from the experimental data.

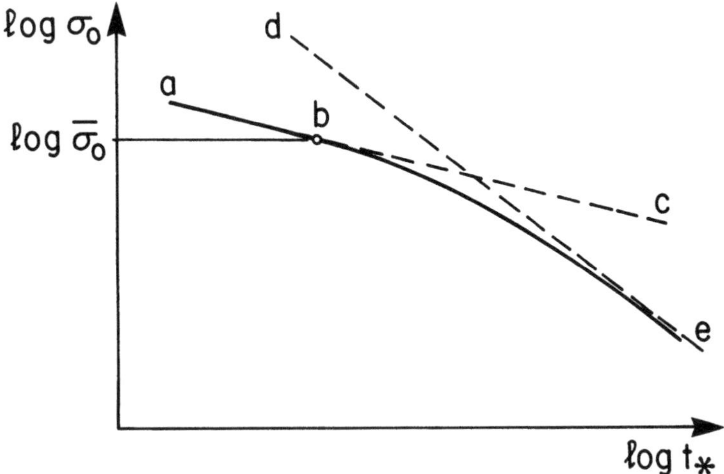

Figure 2.13

(c) *Odqvist's Correction.* Only the creep strain at the stage of steady creep has been considered in the above described model. Odqvist [11] showed that when the first stage of creep is taken into acount, the agreement with the experimental data is better. Odqvist's method can be applied here; we substitute the creep strain accumulated in the first period by instantaneous plastic strain

$$\varepsilon_o = B_o \sigma^{m^o} ,$$

where $B_o \geq 0$ and $m_o \geq 0$ are constants. The rate of the steady creep is described by the equation

$$\dot{\varepsilon}^c = B_1 \sigma^m .$$

Thus, the total strain rate is

$$\dot{\varepsilon} = \frac{d}{dt} B_o \sigma^{m_o} + B_1 \sigma^m . \tag{2.47}$$

According to the experimental data, as a rule, $m > m_o$. Changing the variable in equation (2.47) to $\lambda = \ell / \ell_o$, we obtain, after some transformations, the differential equation

$$(\lambda^{-m-1} - m_o \varepsilon_{oo} \lambda^{-m-1+m_o}) d\lambda = \dot{\varepsilon}_o dt \ , \ (\varepsilon_{oo} = B_o \sigma_o^{m_o}). \tag{2.48}$$

Integration with the initial condition $\lambda = 1$ at $t = 0$ leads to the formula

$$(1-\lambda^{-m})-\beta(1-\lambda^{-m+m_o})=\frac{t}{t_1}\ ;\ t_1=(m\dot{\varepsilon}_o)^{-1} \tag{2.49}$$

where the notation $\beta=mm_o/m-m_o\varepsilon_{oo}\geq 0$ is introduced.

Taking $\lambda \rightarrow \infty$ we obtain the time to viscous fracture,

$$t_{11}=t_1(1-\beta)\ . \tag{2.50}$$

We can see that $t_{11}<t_1$, and this agrees better with the experimental data. In the case of brittle-viscous fracture, the kinetic equation of damage (2.25) and the relation (2.47) must be considered simultaneously.

Integrating the kinetic equation with the initial condition $\psi=1$ at $t=0$ we obtain

$$1-\psi^{n+1}=A(n+1)\int_o^t \sigma^n(\tau)d\tau\ . \tag{2.51}$$

At the moment of fracture t_*, we have $\psi=0$ and the relation (2.51) results in

$$t'=\int_o^{t_*}\lambda^n dt\ ;\ t'=[(n+1)A\sigma_o^n]^{-1}\ .$$

Substituting dt from (2.48) and integrating over λ from $\lambda=1$ to $\lambda=\lambda_*$, we find

$$\frac{t'}{t_1}=\frac{m}{m-n}(1-\lambda_*^{n-m})+\frac{(m-m_o)\beta}{n+m_o-m}(1-\lambda_*^{n-m+m_o}). \tag{2.52}$$

Substitution of λ_* found from (2.52) into (2.49) determines the fracture time t_*.

If $\beta=0$, the formulae (2.52) and (2.49) yield the above found relation (2.44).

Figure 2.14 shows curve 1 of long-time strength according to the solution (2.44) and curve 2 corresponding to the solution with Odquist's correction. Curve 2 is located below curve 1, so the transition to purely viscous fracture occurs earlier (i.e., $\bar{\sigma}_{oo}<\bar{\sigma}_o$). As mentioned before, the agreement with experimental data becomes somewhat better.

Now there are two remarks to be added:

1) The influence of the first period of creep can be approximately described if the constant B_1 in the creep law is substituted by the increasing function of time $B_1\beta(t)$ [9].

2) Let us return to the case of brittle fracture under constant load, section (a). If the creep acceleration in the tertiary period is not great we can introduce the average strain rate

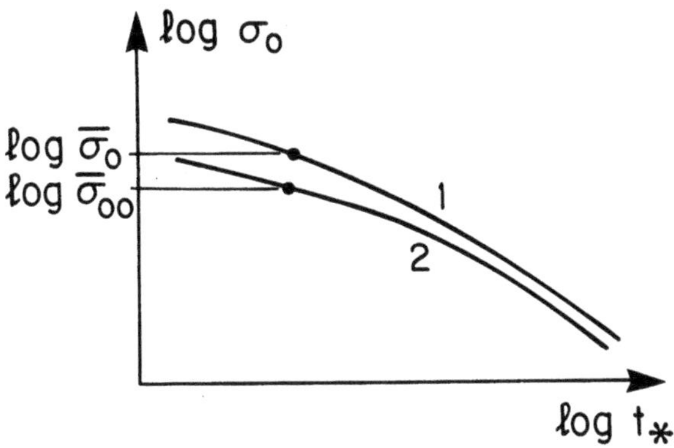

Figure 2.14

$$\dot{\varepsilon}=\frac{\varepsilon'}{t'},$$

where ε' is the strain at fracture. The density of fracture work is

$$w'=\sigma\varepsilon'.$$

The density of the current creep strain work is equal to $w=\sigma\dot{\varepsilon}t$. Thus, the relation (2.28) can also be written in the form

$$\psi=(1-\frac{w}{w'})^{1/n+1}. \tag{2.53}$$

Consequently, the damage parameter ψ is associated with the work of fracture.

(d) *Influence of Damage on Creep Deformation.* It was assumed above that the creep deformation does not depend on the damage parameter. Creep acceleration in the tertiary period, Figure 2.1, cannot be fully attributed to the decrease of the cross-sectional area. Damage resulting from the development of microcracks and microvoids can also lead to creep acceleration.

This effect can be described by introducing (see Rabotnov [12]) the damage parameter ω into the creep equation, i.e.

$$\dot{\varepsilon}=f(\sigma,\omega). \tag{2.54}$$

The kinetic equation has the form

$$\dot{\omega}=g(\sigma,\omega). \tag{2.55}$$

The functions f and g can be found from experimental data by applying a method analogous to that used in technical thermodynamics for the determination of state functions of two variables. For creep problems, the method was described by Leckie and Hayhurst [16].

To simplify the analysis, the power functions are used [12,16]

$$\dot{\varepsilon}=b\sigma^{m}\psi^{-q} , \tag{2.56}$$

$$\dot{\omega}=c\sigma^{n}\psi^{-r} , \tag{2.57}$$

where b, c, m, q, r are material constants, while $m \geq n$. They are determined by the correlation of the solutions of the equations (2.56) and (2.57) (mostly for the case of constant stress) with the experimental data. Note, that sometimes, for convenience, q is assumed to be equal to m; this corresponds in some sense, to the concept of actual stress. Furthermore, it is possible to assume that $b=B_1$.

Consider, at first, the case of purely brittle fracture. Here, the strain is small, and $\sigma = const. = \sigma_o$. Equation (2.57) yields

$$\psi^{r+1}=1-\frac{t}{t'} , \tag{2.58}$$

where $t'=[c(r+1)\sigma_o^n]^{-1}$ is the time to brittle fracture according to the condition $\psi=0$.

Substituting ψ into the creep equation (2.56) and integrating under the initial condition

$$\varepsilon=0 \text{ at } t=0 ,$$

we find that

$$\varepsilon=\frac{\rho}{m}\frac{t'}{t_1}[1-(1-\frac{t}{t'})^{1/\rho}] , \tag{2.59}$$

where $t_1=[bm\sigma_o^m]^{-1}$ is the time to the purely ductile fracture and $\rho=r+1/p$; $p=r+1-q$.

The creep curve according to equation (2.59) is shown in Figure 2.15. We see that by means of this scheme it is possible to describe the tertiary period of creep in the case when the cross-sectional area does not decrease.

Fracture of a mixed type can be considered by solving of the system of the equations (2.56), (2.57) under the condition of finite strain; in this case,

$$\sigma=\sigma_o \exp\varepsilon , \tag{2.60}$$

and the system may be rewritten in the form

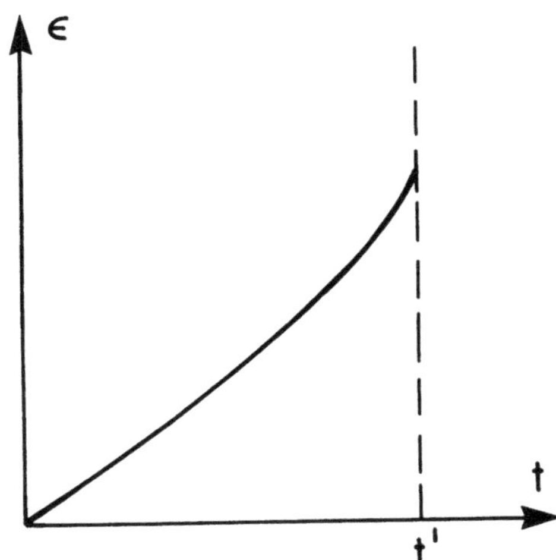

Figure 2.15

$$\frac{d\varepsilon}{dt}=\frac{1}{mt_1}\psi^{-q}exp(m\varepsilon)\ ,$$

$$\frac{d\omega}{dt}=\frac{1}{(r+1)t'}\psi^{-r}\exp(n\varepsilon)\ . \qquad (2.61)$$

Dividing the first equation by the second, we obtain a differential equation with separable variables, i.e.,

$$\frac{d\varepsilon}{d\psi}=-\frac{r+1}{m}\frac{t'}{t_1}\psi^{r-q}\exp[(m-n)\varepsilon]\ .$$

The initial condition is

$$\varepsilon=0 \text{ at } \psi=1\ .$$

It is not difficult to find the corresponding solution, namely,

$$\varepsilon=\frac{1}{m-n}\ell n[1-\nu+\nu\psi^{r-q+1}], m\neq n$$

where

$$\nu=\frac{m-n}{p}\frac{b}{c}\sigma_o^{m-n}\ .$$

In the case $m=n$

$$\psi^p=1-\frac{\varepsilon}{\varepsilon_*}$$

where

$$\varepsilon_* = \frac{b}{cp}.$$

Substituting ψ into the first equation (2.61) and integrating, we obtain

$$\int_0^{\varepsilon} (1 - \frac{\varepsilon}{\varepsilon_*})^{q/p} exp(-m\varepsilon)d\varepsilon = \frac{1}{m}\frac{t}{t_1}.$$

This is the equation of the creep curve. So, the more general scheme does not change the determination of the time to brittle fracture, but makes it possible to describe the tertiary creep period. Therefore, it is possible to consider more completely the stress redistribution in the body and, hence, to estimate more exactly the time of its "life".

Finally, it should be noted that for steels inclined to brittle fracture in long-time tests, the accumulation of creep strain in the tertiary period is sometimes not intensive.

In these cases, the corresponding stress redistribution is not significant.

2.5 Brittle Fracture in Bending of Beams

(a) *Front of Fracture.* If the stress field is uniform (as in extension of a rod), damage increases uniformly, too, until it reaches a critical value at which an instantaneous failure occurs. If the stress field is non-uniform (as, for example, in the bending of a rod), two stages of fracture must be considered. At the first stage (stage of latent fracture), $0 \leq t < t_I$ the continuity ψ is positive at each point of the body. At the moment t_I local fracture appears at a certain point (or region); the scattered microcracks coalesce and macrocracks form. The process of damage accumulation becomes unstable. During the second stage, fracture is characterized, mostly, by the growth of macroscopic cracks.

A rigorous analysis of the development of casual macrocracks is practically impossible. However, for the final stage $t > t_I$ the same scheme of diffused damage can be used, if we introduce the concept of a moving front of fracture.

At the moment $t > t_I$ let fracture spread over the region V_2, Figure 2.16. The region V_2 in the case of bending, is separated from the rest of the body V_1 (where $\psi > 0$) by the moving plane Σ (front of fracture). On the plane Σ $\psi = 0$. Therefore,

$$\frac{d\psi}{dt} = \frac{\partial \psi}{\partial t} + \frac{\partial \psi}{\partial u}\frac{du}{dt}, \tag{2.62}$$

where u is the distance in the direction of propagation of the front.

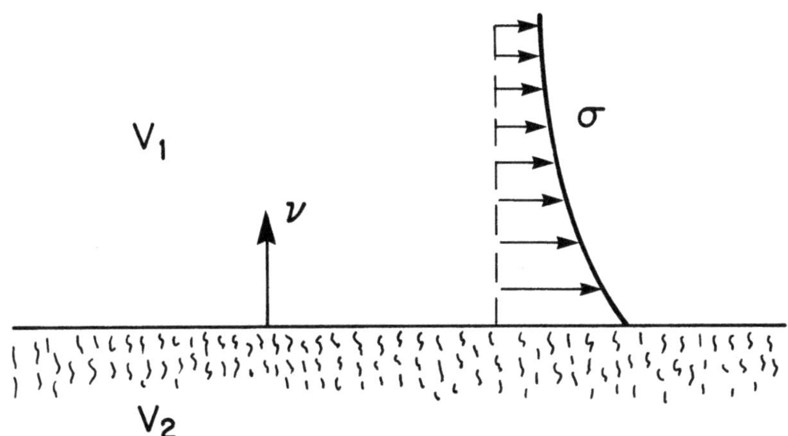

Figure 2.16

Taking into account the kinetic equation (2.25) and the relation (2.51), we obtain from (2.62) the equation of the motion of the fracture front, i.e.,

$$\frac{du}{dt}=(\sigma^n)_\Sigma\left[\frac{\partial}{\partial u}\int_o^t \sigma^n dt\right]^{-1}. \qquad (2.63)$$

the index Σ indicates that the corresponding values are calculated at the front. It is obvious that $t \geq t_I$.

It is also assumed that the stress field does not change rapidly near the front (there is no "boundary effect" for a stress field at the front).

In the case of a uniform stress field, the denominator on the right-hand side becomes zero, so the front of fracture propagates instantaneously. For a non-uniform stress field, the velocity $\dot{u}$ is finite.

The equation of the fracture front can be written in a different form, which is sometimes more convenient.

The stress σ at a certain fixed point of the body at a moment τ is a function of τ and of the coordinates of the point. At the moment t fracture front reaches the mentioned point. Therefore, $\psi=0$ and from the relation (2.51) it follows that

$$A(n+1)\int_o^t \sigma^n(\tau)d\tau=1. \qquad (2.64)$$

Equation (2.63) can be easily derived from equation (2.64).

(b) *The Case of Pure Bending.* Consider the problem of brittle fracture of a beam of a rectangular cross-section in pure bending. We assume that fracture develops at small strain, and the stress distribution is close to a steady one.

At the stage of latent fracture $0 \leq t < t_I$ the continuity $\psi > 0$, so the whole cross-section of the beam resists bending. The normal stress is given by

$$\sigma = \frac{M}{I_{mo}} y_o^{\mu} \ , \ (y_o > 0) \ , \ \mu = \frac{1}{m} \ , \qquad (2.65)$$

where M is the bending moment, x_o, y_o are the coordinates (x_o is directed along the beam axis). The generalized moment of inertia (see, for example, [9]) is

$$I_{mo} = \frac{4b}{2+\mu} h_o^{2+\mu} \qquad (2.66)$$

where $2b$ and $2h_o$ are the width and the height of the cross-section, correspondingly.

In the stretched zone $y_o > 0$ the continuity ψ is determined by the kinetic equation (2.25), $\psi = 1$ and there is no fracture in the compressed zone $y_o < 0$. The stress σ is maximal at $y_o = h_o$ and we obtain from formula (2.51) that

$$t_I = [(n+1)A(\frac{M}{I_{mo}})^n h_o^{n/m}]^{-1} \ . \qquad (2.67)$$

(c) *Motion of the Front of Fracture.* At the moment t_I the surface layer $y_o = h_o$ fractures. Then, the front of fracture Σ moves into the beam, Figure 2.17. Let the thickness of the fractured layer be 2δ at the moment t; it is obvious that $h_o = h + \delta$. At a certain moment $\tau > t_1$ the stress will be

$$\sigma = \frac{M}{I_m} y^{\mu} \ , \qquad (2.68)$$

where the generalized moment of inertia

$$I_m = \frac{4b}{2+\mu} h^{2+\mu}$$

is a function of τ, and y is reckoned from the new position of the neutral axis, while $y = y_o + h_o - h$. Let the front of fracture at the moment t pass through a fixed point with the coordinate y_o. According to the equation (2.64) we have

$$(n+1)AM^n \int_o^t [2h(t) - h(\tau)]^{n/m} \frac{d\tau}{I_m^n(\tau)} = 1 \ .$$

But at this moment of time $y = h(t)$. Therefore, $y_o + h_o = 2h(t)$ and the latter equation becomes

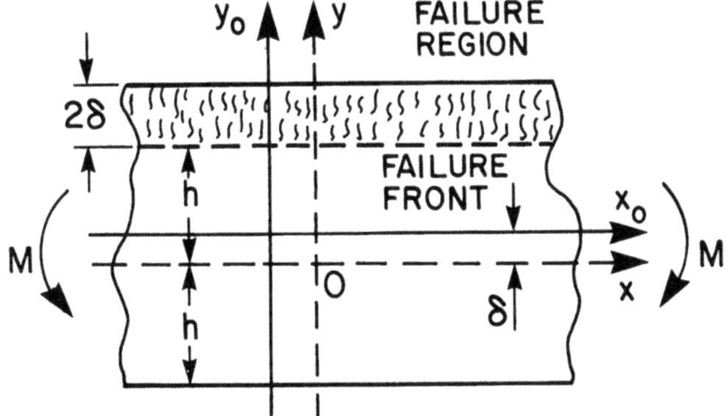

Figure 2.17

$$(n+1)AM^n\int_o^t [2h(t)-h(\tau)]^{n/m}\frac{d\tau}{I_m^n(\tau)}=1 .$$

Assume, for simplicity, that $m=n$; differentiating the equation with respect to t, we obtain

$$2\frac{dh}{dt}\int_o^t h^{-1-2m}(\tau)d\tau+h^{-2m}=0. \tag{2.69}$$

The initial condition is

$$h=h_o \text{ at } t=t_I .$$

We also find from (2.69), that

$$\frac{dh}{dt}=-\frac{h_o}{2t_I} \text{ at } t=t_I .$$

Differentiating equation (2.69) with respect to time, and eliminating the integral term, we obtain the differential equation for $h(t)$

$$\frac{d^2h}{dt^2}+2(m-1)\frac{1}{h}(\frac{dh}{dt})^2=0.$$

Integrating this equation and finding the arbitrary constants from the initial conditions, we obtain

$$\frac{t}{t_I}=1+\frac{2}{2m-1}[1-(\frac{h}{h_o})^{2m-1}]. \tag{2.70}$$

At $h=0$ the beam is completely fractured; the corresponding time of fracture is

$$\frac{t'}{t_I}=1+\frac{2}{2m-1}\ .$$

Hence, the stage of propagation of fracture is relatively long (for instance, $t'=1.4t_I$ in the case $m=3$).

(d) *Fracture of a Beam under Arbitrary Load.* Following the usual scheme of the technical theory of bending, we can apply the obtained results to the analysis of fracture of a beam under arbitrary load. The stress σ will be determined by the formula (2.65) mentioned before, but the bending moment will be a function of x: $M=M(x)$.

Let the bending moment be maximal at $x=x^*$. Fracture occurs in the surface layer of the extended zone at the point x^* at the time moment t_I^*; the latter is determined by the formula (2.67), where $M=M(x^*)$. The zone of fracture will spread over the entire cross-section $x=x^*$ at the moment

$$t'_*=t_I^*(1+\frac{2}{2m-1})\ . \tag{2.71}$$

At the moment of time t'_* fracture will spread over a certain wedge-like region adjacent to the extended side of the beam in the vicinity of the cross-section $x=x^*$. The boundary of this zone $h=h(x)$ can be found from the formula (2.70) if we take

$$t=t'_*\ ;\ t_I=t_I(x)$$

where $t_I(x)$ is the time of latent fracture of the beam's surface layer in the cross-section x. The equations (2.67) and (2.71) yield

$$[\frac{M(x)}{M(x^*)}]^m=\frac{2m-1}{2m+1}\{1+\frac{2}{2m-1}[1-(\frac{h(x)}{h_o})^{2m-1}]\}\ . \tag{2.72}$$

Prescribing the values $h(x)/h_o$, we can find the corresponding coordinates x from (2.72).

Consider, for example, the fracture of a beam of the length 2ℓ resting on two supports and bent by the force P applied in the middle of the beam; then,

$$\frac{M(x)}{M(x^*)}=1-\frac{x}{\ell}\ .$$

Fracture spreads over a narrow region with an edge, Figure 2.18.

2.6 Brittle Fracture in Cyclic Bending

Consider a simple case of pure bending of a beam of rectangular cross-section with width 2b, Figure 2.17. The cycle is symmetric ($+M$, $-M$, $+M\ \cdots$), its steps are of equal length.

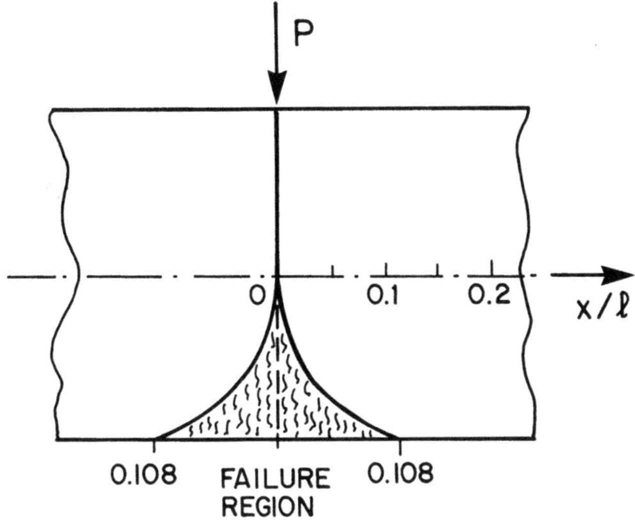

Figure 2.18

In cyclic loading there arises the problem of resistance of the fractured zone to compression (beyond the front Σ). The assumption that the fractured zone offers no resistance to compression is the safest; it can be accounted for by the fact that the crack's edges gradually close at the first stage of compression. Then, when the cracks close completely, the resistance is restored to some extent. It is also expedient to consider another extreme case when the fractured zone does resist compression.

The steps are of equal length. Since a sufficiently great number of cycles has been realized by the moment of fracture, displacements of the neutral axis can be neglected. Therefore, this axis is assumed to be identical with the initial one.

For solving this problem, the stress distribution under creep must be found. The problem of the resistance of fractured zones to creep arises here again. Since the creep strains grow slowly, and the period T is not too long this resistance is assumed not to exist; only the core $|y| \leq h$ offers resistance. In the zones $|y| > h$ the stresses are zero. The stress distribution in the core (for $y > 0$) is given by

$$\sigma = \frac{M}{I_m} y^{\mu} \; ; \; I_m = \frac{4b}{2+\mu} h^{2+\mu}$$

We also assume that changes in the loading regime do not influence the stress significantly, which can be attributed to the slow character of such changes. Hence, this quasistationary distribution is realized in the low-frequency regime. At the stage of latent fracture $t<t_1$ we have $h=h_o$. The time t_I is given by

$$t_I=2[(n+1)AM^n(\frac{2+\mu}{4bh_o^2})^n]^{-1}\equiv 2Dh_o^{2n}\,.$$

At $t>t_I$ two fracture fronts move, alternately, from the periphery to the axis of the beam. At the front we have the condition

$$1=\frac{1}{D}\int_o^t h^{n/m}(t)[h^{2n+n/m}(\tau)]^{-1}d\tau\,.$$

The time thus found must be multiplied by 2, since fracture does not occur in compression. Differentation of the latter equation with respect to time and elimination of the integral term yields the differential equation

$$D\frac{n}{m}\frac{dh}{dt}+h^{1-2n}=0$$

whose solution satisfying the initial condition $h=h_o$ at $t=t_I$ is

$$1-(\frac{h}{h_o})^{2n}=\frac{2m}{Dh_o^{2n}}(t-t_I)\,.$$

Assuming $h\to 0$ and multiplying the result by 2 we find the time to fracture:

$$t'=t_I(1+\frac{1}{2m})\,.$$

In the case of constant moment ($M=const.$), the time to fracture (sect. 2.5, (c)) is

$$t'=\bar{t}_1\frac{2m+1}{2m-1}\,;\ \bar{t}_1=\frac{1}{2}t_I\,.$$

Consequently, the relative time of front propagation is significantly shorter under the cyclic loading.

The dependence of the ratio $t'/\bar{t}'$ on the creep index m is shown in Figure 2.19. At $m\to\infty$ the stress distribution approaches an ideally plastic one; then $t'\to 2\bar{t}'$.

Consider now the second extreme case when the fractured part of the cross-section does not resist extension but strongly resists compression. Then, in the case $m=n$ the motion of the fracture front is described by the differential equation

$$\frac{d^2h_1}{dt^2}+2(m-1)\frac{1}{h_1}(\frac{dh_1}{dt})^2=0;(2h_1=h_o+h)$$

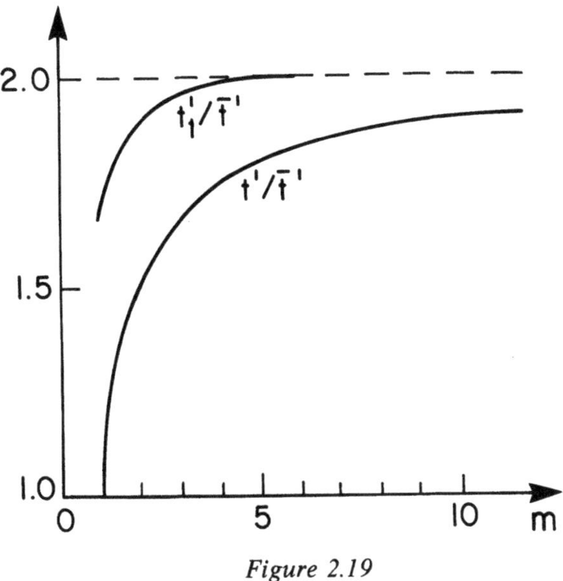

Figure 2.19

which has already been considered in the analysis of the fracture of a beam loaded by a constant moment. Fracture occurs at $h_1 \rightarrow 1/2h_o$. The corresponding moment of time t'_1 is given by

$$\frac{t'_1}{\bar{t}'} = 2\{1 - [2^m(2m+1)]^{-1}\} .$$

This dependence is shown in Figure 2.19. If the fractured zone does not resist compression, the time to fracture is significantly shorter.

Within the scope of the technical theory of the bending of beams, the obtained results can be used for the analysis of the fracture of a beam under an arbitrary symmetric cyclic loading with steps of equal length. The bending moment M will be a function of x. Note, in conclusion, that various cross-sections of beams and various cycles can be considered in an analogous way.

2.7 Fracture of a Rotating Shaft in Bending

Consider the fracture of a slowly rotating circular shaft (radius a) in pure bending. If the rotation is slow the stress redistribution can be neglected and the stresses can be assumed to be the same as in the state of steady creep.

For sufficiently great number of cycles, damage can be treated as axially symmetric. A ring-shaped zone of full fracture $r \geq c$ forms at $t > t_I$. The radius of the unfractured core $c(t)$ decreases with time. Assume that the fractured zone does not resist compression. Then the bending moment is

$$M = M_o \sin\omega t \quad (M_o > 0)$$

and the normal stress is

$$\sigma = \frac{M}{I_m} y^{\mu}$$

where $I_m = q(\mu) c^{3+\mu}$ is the generalized moment of inertia; the values $q(\mu)$ are given, for example, in [9].

Let us find the time t_I. In the stage of latent fracture $0 \leq t \leq t_I$ we have $c = a$, $\sigma = M / q(\mu) a^{-3}$ and a symmetric harmonic cycle takes place. Therefore, $t_I = \kappa T$, where the multiplier κ is determined by the formula (2.38) with t'_o being the time to fracture at constant stress σ when $M = M_o$.

Now, consider the motion of the fracture front at $t > t_I$; it is determined by the equation

$$1 = (n+1) A \int_o^t \left[\frac{M(\tau)}{q(\mu) c^3(\tau)} \left(\frac{c(t)}{c(\tau)}\right)^{\mu} H(\tau)\right]^n d\tau$$

where $H(\tau) = 1$ for $\sigma > 0$ and equals zero for $\sigma < 0$.

Differentiating this relation with respect to t and eliminating the integral term, we obtain

$$\frac{dc}{dt} + N c^{1-3n} H(t) sin^n \omega t = 0$$

where

$$N = \frac{m}{n} \frac{(n+1)A}{q^n(\mu)} M_o^n > 0 .$$

It is not difficult to see that $dc/dt < 0$. The initial condition is $c = a$ at $t = t_I$. The solution of the problem is given by

$$c^{3n} - a^{3n} = -3nKN(t - t_I) .$$

Assuming $c \to 0$, we obtain the time to fracture as

$$t' = t_I \left(1 - \frac{1}{3m}\right) .$$

Thus, the time of propagation of the fracture front is relatively short.

2.8 Fracture of a Non-Uniformly Heated Flat Wall under Tension

(a) *Introduction.* Time to brittle fracture is considerably affected by the temperature θ. It has been previously noted (sect. 2.3) that if the interval of temperatures is not too large, then in the kinetic equation (2.25) the index n can be considered as constant and the coefficient A as an exponential function of temperature, i.e.

$$A = A_o e^{\alpha\theta} ,$$

where A_o and α are constants.

A flat non-uniformly heated wall is a commonly used heat-transferring element of various machines and structures. The heat-transferring thin shells (tubes, bottoms, etc.) can locally be considered as flat walls. Consider an example of brittle fracture of a stretched non-uniformly heated wall [10].

(b) *Fracture Time of a Non-Uniformly Heated Wall under Tension.* The wall is stretched by forces P in the y-direction, Figure 2.20. The temperature of the wall $\theta = \theta(x)$ is stationary and monotonically decreasing in direction of negative x. Under the conditions of plane strain, the strain rate $\dot{\varepsilon}_z = 0$. Then, it follows from the equations of steady creep that the mean stress $\sigma = 1/2(\sigma_x + \sigma_y)$. Considering, for simplicity, the case of a power law, we have

$$\dot{\varepsilon}_x = -\dot{\varepsilon}_y = \frac{1}{4} B T^{m-1}(\sigma_x - \sigma_y) , \tag{2.73}$$

where B and m are constants. Since the wall is long, the stresses and velocities are functions of x alone. Then, it follows from the equilibrium condition that $\sigma_x = 0$; the intensity of the shear stresses is $T = 1/2 \mid \sigma_y \mid$, and the relation (2.73) becomes

$$\dot{\varepsilon}_x = -\dot{\varepsilon}_y = -\frac{B}{2} \left| \frac{\sigma_y}{2} \right|^{m-1} \frac{\sigma_y}{2} . \tag{2.74}$$

Neglecting a possible warping of the wall (it can decrease in stretching), we obtain $\dot{\varepsilon}_y = const. = C_1$; we assume here $\sigma_y > 0$. We can assume (see [9]) that the index m does not depend on temperature, and

$$B = B(\theta) = B_o e^{\beta\theta}$$

where β and B_o are constants. The stress σ_y is easily found from (2.74)

$$\sigma_y = K \exp(-\frac{\beta}{m}\theta) \ ; \ K = 2(\frac{2C_1}{B_o})^{\mu} . \tag{2.75}$$

Determining C_1 from the condition of equilibrium, we find

$$K = P[\int_{-h}^{h} \exp(-\frac{\beta}{m}\theta) dx]^{-1} . \tag{2.76}$$

For a steady heat flux,

$$\theta(x)=a+bx\ ;\ a=\frac{1}{2}(\theta_1+\theta_2)\ ;\ b=\frac{1}{2h}(\theta_1-\theta_2)$$

where θ_1,θ_2 are the temperatures of the boundaries $x=h$, $x=-h$, respectively.

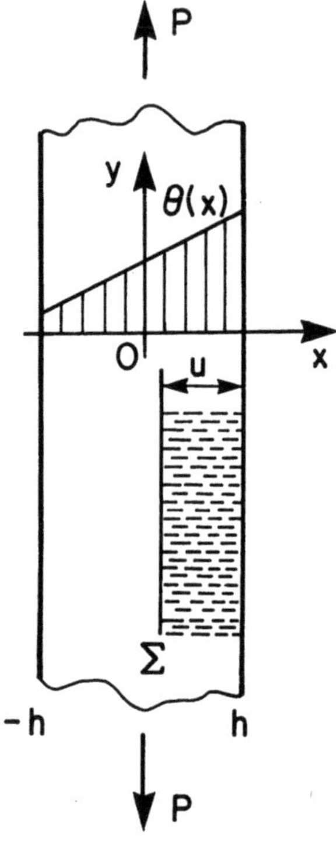

Figure 2.20

At the first stage of fracture (prior to the formation of the fracture front), the stress σ_y does not depend on time. Then, the kinetic equation for damage yields

$$\psi^{n+1}=1-(n+1)A_oK^n e^{\kappa\theta}t \tag{2.77}$$

where $\kappa=\alpha-\beta n/m$. Depending on the sign of κ, the intensity of fracture is higher either on the front side $x=h$ for $\kappa>0$ or on the back side $x=-h$ for $\kappa<0$.

In the case that creep does not depend on the temperature, $\beta=0$ and $\kappa>0$.

If the process of damage accumulation is temperature-independent, then $\alpha=0$ and $\kappa<0$.

The time of latent fracture t_I is to be determined from (2.75) for $\psi=0$ and for $x=h\cdot sign\kappa$. For definiteness, let $\kappa>0$. Then, the fracture front propagates from the boundary $x=h$. On the front Σ we have the condition

$$(n+1)\int_o^t A\sigma_y^n(\tau)d\tau=1\,. \tag{2.78}$$

Let the front pass through a point with the coordinate $h-v$ at the moment t and through a point $h-u$ at the moment $\tau<t$, Figure 2.20.

Determining σ_y at the point $h-v$ at the moment τ from (2.75) and (2.76), introducing it into (2.78), and taking $v=u$, we obtain

$$(n+1)A_oP^n e^{\kappa\theta(h-u)}\int_o^t [\int_{-h}^{h-u(\tau)} \exp(-\frac{\beta}{m}\theta)dx]^{-n}d\tau=1\,.$$

Differentiating with respect to t and eliminating the integral term, we obtain the differential equation

$$e^{-\kappa\theta(h-u)}\theta'(h-u)[\int_{-h}^{h-u} \exp(-\frac{\beta}{m}\theta)dx]^n du=\frac{n+1}{\kappa}A_oP^n dt$$

where $\theta'(h-u)>0$ denotes the derivative.

Introducing θ and performing the integration with the initial condition $u=0$ for $t=t_I$, we obtain for $b\neq 0$, $\beta\neq 0$,

$$\kappa b(\frac{m}{\beta b})^n e^{-\alpha\theta_1+\frac{\beta n}{m}(\theta_1-\theta_2)}\int_o^u e^{\kappa bu}\{1-exp[-\frac{\beta b}{m}(2h-u)]\}^n du=$$
$$=(n+1)A_oP^n(t-t_I)\,.$$

The integral is easily calculated when n is an integer. Assuming $u=2h$, we find the time to fracture as

$$t'=t_I+t'_oI_n\,, \tag{2.79}$$

where I_n is given by a certain sum and

$$t'_0=[(n+1)A_0e^{n/2(\theta_1+\theta_2)}(\frac{P}{2h})^n]^{-1}\,.$$

Here, t'_o is the time to fracture of the uniformly heated wall the temperature of which is $1/2(\theta_1+\theta_2)$. If $\beta=0$ formula (2.79) is also easily obtained. But, in this case,

$$I_n=qe^{1/2q}\int_o^1 e^{-q\xi}\xi^n d\xi\;;\; q=\alpha(\theta_1-\theta_2)$$

and the formula (2.79) can be rewritten in the form

$$\frac{t'}{t_I} = 1 + e^{q/2} I_n \, .$$

The graphs of the second term for $n=1$, $n=3$ are given in the Figure 2.21. It should be noted that here, unlike as in the case of a uniformly-heated body, the relative time of propagation of the front can be long. It is explained by a sharp decrease of the rate of fracture with decreasing temperature.

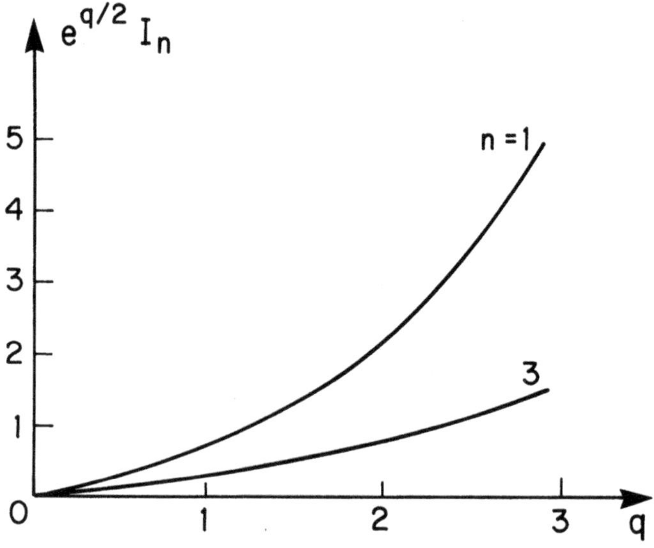

Figure 2.21

(c) *Note:* Both the problem of brittle fracture of a non-uniformly heated wall under the conditions of pure bending in x,y plane and the problem of a wall stretched in two directions y and z have been considered [10].

2.9 Influence of an Aggressive Medium on Brittle Fracture

(a) *Influence of Diffusion of some Elements on Fracture.* Diffusion of some elements can significantly reduce the long-term resistance of steels; it depends on the gradient of their concentration and, generally, on the level of stress. We consider here the case of a liquid metal medium (used, for example, in nuclear energetics). In this case, as a rule, no significant influence of the stress state on diffusion is observed. Thus, the process of diffusion can be considered as independent of the stress field. But the fracture process depends essentially on the stress state and on the concentration of the element in the steel.

Consider the fracture of a circular rod (with the initial diameter $2a_o$) loaded by a tensile force P.

Let $q=q(r,t)$ be the concentration of the element, the initial concentration being

$$q(r,0)=q_o<0 \ , \ (0\leq r\leq a_o) \ .$$

We assume, for simplicity, that the concentration of the element in the aggressive medium is equal to zero. The concentration $q(r,t)$ is determined by the differential equation

$$\frac{\partial q}{\partial t}=\beta(\frac{\partial^2 q}{\partial r^2}+\frac{1}{r}\frac{\partial q}{\partial r}) \ ,$$

where β is the diffusion coefficient. We use the following classical boundary condition

$$\frac{\partial q}{\partial r}=-\gamma q \text{ at } r=a_o$$

where γ is some coefficient.

The solution of this classical problem by Fourier's method is well-known. For illustration, the concentration of the element as a function of the radius r for two moments of time $t_2>t_1>0$ is shown in Figure 2.22. We assume that the process of brittle fracture can be described by the same equation we used above, i.e.,

$$\frac{d\psi}{dt}=-A(\frac{\sigma}{\psi})^n \ ,$$

with the coefficient A depending on the concentration q.

Let this dependence be linear:

$$A=A_o-A_1(q_o-q)$$

where $A_o>0$ and A_1 are some coefficients. With the development of diffusion the fracture process accelerates if $A_1>0$.

Separating variables in the kinetic equation and integrating, we obtain

$$\psi^{n+1}=1-\frac{1}{t'_o}[(1-\alpha q_o)\int_o^t s^n dt+\alpha\int_o^t q s^n dt] \ , \tag{2.80}$$

where

$$s=(\frac{a_o}{a})^2 \ , \ \alpha=\frac{A_1}{A_o} \ , \ \sigma=s\sigma_o$$

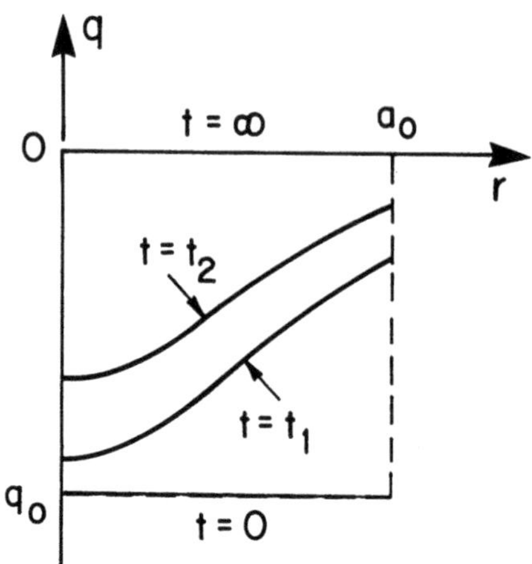

Figure 2.22

$$\sigma_o = \frac{P}{\pi a_o^2}\ ;\ t'_o = [(n+1)A\sigma_o^n]^{-1}\ .$$

t'_o is the time to brittle fracture in the absence of diffusion, $2a$ is the diameter of the non-fractured core of the rod.

At the stage of latent fracture $t \leq t_I$ and when the strain is small, $a = a_o$, and it follows from (2.80) that

$$t'_o - (1-\alpha q_o)t_1 - \alpha\int_o^{t_I} q\,dt = 0\ .$$

This relation determines the time t_I ($\psi = 0$ at $t = t_I$ and $r = a_o$). At $\alpha = 0$, $t_I = t'_o$ and at this moment the entire cross-section fractures.

Consider now the stage $t > t_I$. The concentration changes most at the boundary $a = a_o$. At the moment $t = t_I$, the surface layer of the rod fractures, and there forms a fracture front $a(t)$.

Assume, for simplicity, that diffusion is not influenced by the process of fracture.

On the fracture front $r = a$ we have $\psi = 0$. Therefore, $d\psi/dt = 0$. So,

$$\frac{da}{dt} = -\left(\frac{\partial\psi}{\partial t}\right)_a\left(\frac{\partial\psi}{\partial r}\right)_a^{-1}\ .$$

Now, using (2.80), it is not difficult to find that for $t > t_I$ we have

$$\frac{da}{dt} = -[1-\alpha(q_o - q)]s^n\left[\alpha\int_o^t\left(\frac{\partial q}{\partial r}\right)_a s^n\,dt\right]^{-1}\ . \qquad (2.81)$$

At sufficiently large t the difference $(q_o - q)_a \approx q_o$. Differentiation of (2.81) with respect to time yields

$$\frac{d^2a}{dt^2}\left(\frac{da}{dt}\right)^{-2} + \frac{2n}{a} = \frac{\alpha}{1-\alpha q_o}\left(\frac{\partial a}{\partial r}\right)_a .$$

In the general case, this equation is to be integrated numerically. If the time t_I is sufficiently long (so that the concentration q becomes small) the value of the derivative in the right-hand side of the latter equation will be small. Hence,

$$\frac{d^2a}{dt^2} + \frac{2n}{a}\left(\frac{da}{dt}\right)^2 \approx 0 .$$

Integrating this equation we obtain

$$\frac{da}{dt} = -D_1 a^{-2n} \tag{2.82}$$

where D_1 is an arbitrary constant.

The value of the derivative $(da/dt)_{t_I}$ (denoted by $\dot{a}_I$) can be obtained from equation (2.81) when $t = t_I$, $a = a_o$, $s = 1$. Then, the equation (2.82) will have the form

$$\frac{da}{dt} = \dot{a}_I\left(\frac{a_o}{a}\right)^{2n} .$$

Integrating once more and determining the arbitrary constant from the initial condition $a = a_o$ at $t = t_I$, we obtain

$$\left(\frac{a}{a_o}\right)^{2n+1} = 1 + \frac{\dot{a}_I}{a_o}(1+2n)(t - t_I) .$$

At the moment t'' of the complete fracture of the rod, $a = 0$. Thus,

$$t'' = t_I - \frac{1}{1+2n}\frac{a_o}{\dot{a}_I} .$$

Note that $\dot{a}_I < 0$. The duration of the stage of front propagation depends on the characteristics of diffusion and on the parameter α.

The obtained solution makes it possible to estimate the scale effect for the fracture of rods of different diameters; this effect depends on the rod size affecting the process of diffusion.

(b) *Influence of Dissolution on Fracture.* The dissolution of the surface layer of a rod is one of the forms of corrosion and depends, generally, on the level of stress.

But in the case when a steel rod is surrounded by any other metallic medium and no electrochemical processes take place, the influence of stress on the process of dissolution is insignificant and can be neglected.

Consider the problem of the extension of a circular rod in a liquid-metal medium. Let the radius of the rod be a_o at the initial moment of time and $a=a(t)$ at the moment t. The rate of surface layer dissolution $\dot{a}$ depends on the saturation of the medium. If the medium volume is sufficiently large and an intensive mixing takes place there, $\dot{a}$ can be assumed as constant

$$\frac{da}{dt}=const. =-k .$$

Hence

$$a=a_o(1-\frac{k}{a_o}t) ; a\geq 0 .$$

If the change in the saturation of the medium cannot be neglected , $\dot{a}$ is a variable. In this case, the current size of the rod can be found in the form $a=a_o\phi(t)$, $\phi(t)$, $\phi(0)=1$. At the moment $t=t_d$ $\phi(t_d)=0$, i.e., the rod completely dissolves. If k is constant, $t_d=a_o/k$.

Consider now the process of brittle fracture of a slowly dissolving rod. Consider the kinetic equation (2.25) with the coefficient A being constant. It is evident, that

$$\sigma=\sigma_o\phi^{-2}(t) .$$

Assume that

$$\int_o^t\phi^{-2n}(t)dt=\Phi(t) .$$

Then, it follows from equation (2.25) that

$$\psi^{n+1}=1-\frac{1}{t'_o}\Phi(t) ,$$

where t'_o is the time to brittle fracture in the absence of "corrosion". The rod will be fractured at the moment $t=t'$ when $\psi=0$. Hence,

$$\Phi(t')=t'_o ,$$

and the fracture time is

$$t'=\Phi^{-1}(t'_o) ,$$

where Φ^{-1} denotes the inverse function. The solution is valid if $t'<t_d$. This condition is satisfied. Indeed, let $\dot{a}=a_o\phi'(t_d)\neq 0$. Then, since $n>1$, the function $\Phi(t)\rightarrow\infty$ as $t\rightarrow t_d$. Hence, the root $t'<t_d$ of the equation (2.83) exists. In the special case, when k is constant, we have

$$\frac{t'}{t_d}=1-[1+(2n-1)\frac{t'_o}{t_d}]^{-1/2n-1}.$$

It is not difficult to see that the decrease of the long term strength may be significant even in the case of slow dissolution if the index n is sufficiently large.

2.10 Fracture of an Adhesive Bond in Bending

(a) *Bending of a Strip.* Consider the plane stress problem of gradual delamination in bending of an elastic strip bonded to a rigid plane, Figure 2.23. At the initial moment of time $t=0$ the thin adhesive layer ("adhesive line") is fractured along the segment $(0, \ell_o)$. The strip of thickness h is elastic (Young's modulus will be denoted by E). The adhesive layer of thickness δ is viscoelastic, its quasistatic response being elastic. Since the adhesive layer is thin, it can be assumed that there is only the normal stress $\sigma_y \equiv q$ in it. Therefore, the problem can be represented as a strip on an elastic foundation, bent by a force P (per unit width). The equation of bending is

$$EI\frac{d^4v}{dx^4}=q\ , \tag{2.84}$$

where v is the deflection, I is the moment of inertia, and the coordinate x is reckoned from the current position of the fracture front Σ. The vertical displacement of the layer on the contact line is v, while the elastic response is $-E_1/\delta v$, where E_1 is the "elastic modulus" of the layer. Hence the equation (2.84) takes the form

$$v^{IV}+4k^4v=0\ , \tag{2.85}$$

where $4k^4=E_1/EI\delta$. The bond is sufficiently long, and we can assume that

$$v\rightarrow 0 \text{ as } x\rightarrow\infty\ .$$

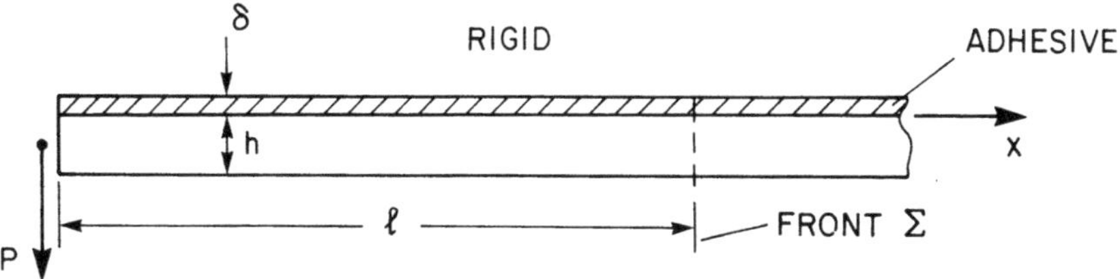

Figure 2.23

Under this condition, the solution of the differential equation (2.85) has the form

$$v = C_1 Z_1 + C_2 Z_2 \tag{2.86}$$

where C_1, C_2 are arbitrary constants and

$$Z_1 = e^{-kx} sinkx \; ; \; Z_2 = e^{-kx} coskx \; .$$

The boundary conditions are:

$$v'' = \frac{P\ell}{EI} \; ; \; v''' = -\frac{P}{EI} \text{ at } x = 0 \; .$$

Then, we obtain

$$C_1 = \frac{P\ell}{2k^2 EI} \; ; \; C_2 = -C_1 (1 - \frac{1}{k\ell}) \; .$$

Note that

$$\frac{1}{k\ell} = (\frac{E}{E_1} \frac{h^3 \delta}{12\ell^4})^{1/2} << 1 \; .$$

Hence, the displacement at $x = 0$ is

$$v \,|_o \approx \frac{P\ell}{2k^2 EI} \; .$$

Let the adhesive layer be fractured along the segment $(0,\ell)$ at the moment of time t. Consider the propagation of the fracture front Σ. According to the obtained solution, the displacement v_m and the stress σ_m at the front Σ are, respectively,

$$v_m = \frac{P\ell}{2k^2 EI} \; ; \; \sigma_m = E_1 \frac{v_m}{\delta} \; .$$

The displacement v_m is determined by (2.86) with $x = 0$. Thus,

$$\sigma_m = \kappa \ell$$

where

$$\kappa = \frac{E_1}{E} \frac{P}{2k^2 I \delta} \; .$$

Under the action of the tensile stress σ_m the layer at the front Σ is creeping and damage develops. We assume that damage accumulation can be described by the simple kinetic equation (2.25), as before.

Hence, after loading, the element located at the distance ℓ_o will be fractured at the moment of time

$$t_1=[(n+1)A\sigma^n]^{-1}$$

where the stress is $\sigma=\kappa\ell_o$. Then the fracture front will start to propagate. The equation of propagation of the fracture front, in general, has the form (2.63). In the problem under consideration, u must be replaced by x, du/dt by $d\ell/dt$; the stress σ has the maximum σ_m for $x=0$ and is characterized, according to the solution (2.86) by the boundary effect at the front Σ.

Therefore, the derivative

$$\left(\frac{\partial}{\partial x}\int_o^t \sigma^n dt\right)_\Sigma$$

in equation (2.63) will misrepresent the analysis and its average value is to be introduced. If, as a first approximation, the denominator in (2.63) is assumed to be constant, we have

$$\frac{d\ell}{dt}=\lambda\sigma_m^n , \tag{2.87}$$

where the coefficient λ is to be found experimentally. Note that the result (2.87) can also be obtained in a different way. Thus,

$$\frac{d\ell}{dt}=c\ell^n ,$$

where $c=\lambda\kappa^n$. The initial condition is

$$\ell=\ell_o \text{ at } t=t_1 .$$

The solution is

$$s=[\alpha(\tau_*-\tau)]^{-1/n-1} ; \ \tau\leq\tau_*$$

where

$$s=\frac{\ell}{\ell_o}\geq 1 , \ \tau=\frac{t}{t_I}\geq 1 ,$$

$$\alpha=(n-1)ct_I\ell_o ; \ \tau_*=1+\frac{1}{\alpha} .$$

According to this solution, fracture develops with acceleration. Total fracture occurs at the finite moment of time τ_*. If $\tau\to\tau_*$, then $s\to\infty$, $ds/d\tau\to\infty$; if $\tau\to 1$, then, $s\to 1$, $ds/d\tau\to\alpha/n+1>0$, Figure 2.24.

The rate of delamination depends on the type of loading. If, for example, fracture is caused by a constant bending moment, then the length ℓ is proportional to time t.

(b) *"Blister Test" (Linear Approach)*. In this case, an elastic plate (of thickness h) is bonded to a rigid plane, except for a central portion of radius b_o at the initial moment of time $t=0$. Under the uniform pressure

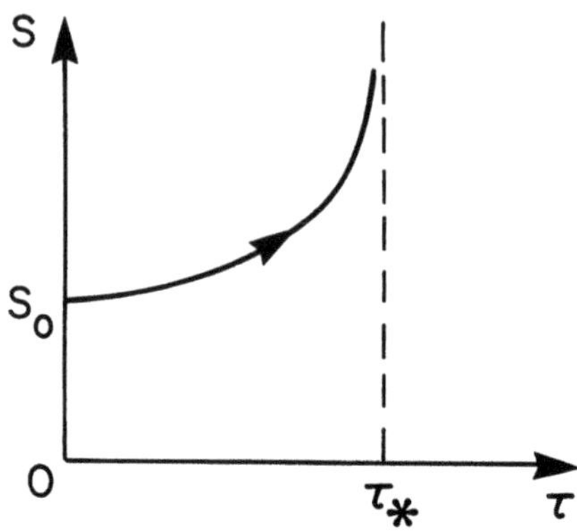

Figure 2.24

p the radius of the "blister" increases. Let the radius be equal to b at time t.

Since the adhesive layer is thin (thickness δ) it is possible to assume that the main stress in it is the stress σ normal to the layer. Treating the adhesive layer as above, we consider an axisymmetrical bending of a plate on an elastic foundation. The solution of the corresponding linear differential equation for the deflection v has the form [44]:

$$w = C_1 bei\, x + C_2 ber\, x + C_3 kei\, x + C_4 ker\, x$$

where

$$w = \frac{v}{\ell}\ ;\ x = \frac{r}{\ell}\ ;\ \ell^4 = \frac{D}{k_1}\ ;$$

$$\frac{b}{\ell} = a\ ;\ k_1 = \frac{E_1}{\delta}\ .$$

Here, D is the flexural rigidity of the plate, r radial distance; C_1, C_2, C_3, C_4 arbitrary constants; $bei\, x$, $ber\, x$, $kei\, x$, $ker\, x$, are Bessel functions. We can assume that $w \to 0$ as $x \to \infty$. Then, $C_1 = C_2 = 0$. For $x = a$ we have the condition of opening:

$$\frac{dw}{dx} = 0\ ,$$

and, besides, the value of the shearing force Q_r is specified, i.e.,

$$Q_r = \frac{1}{2} p a \ell\ .$$

Let us denote the values of the functions *kei* x and *ker* x at $x=a$ by α and β respectively; the values of the derivatives of these functions at $x=a$ by $\alpha_1, \alpha_2, \ldots; \beta_1, \beta_2, \ldots$, respectively, where the index numbers correspond to the order of derivatives.

Using the boundary conditions at $x=a$ we find

$$C_3=\frac{pa\ell^3}{\mu D}, \quad C_4=\frac{\alpha_1}{\beta_1}C_3$$

where

$$\mu=(\alpha_3+\frac{1}{a}\alpha_2-\frac{1}{a^2}\alpha_1)-\frac{\alpha_1}{\beta_1}(\beta_3+\frac{1}{a}\beta_2-\frac{1}{a^2}\beta_1)$$

The tensile stress σ_m at the front $x=a$ is

$$\sigma_m=\frac{E_1}{\delta}\ell w\,|_{x=a}=\frac{E_i\ell}{\delta}(C_3\alpha+C_4\beta)\equiv S(a)\,.$$

Using the same equation (2.87) we obtain the differential equation for the velocity of the fracture front as being

$$\frac{da}{dt}=c_1S^n(a)\,,$$

where $c_1=\lambda/\ell$. The initial condition is

$$a=a_o=\frac{b_o}{\ell} \text{ at } t=t_1\,.$$

Thus, the solution has the form

$$V(a),=c_1(t-t_I)\,,$$

where

$$V(a)=\int_{a_o}^{a}S^{-n}(a)da\,.$$

The function $V(a)$ can be found numerically.

Note that $a>>1$. Hence, it is possible to use asymptotic expressions for Bessel functions.

(c) *Concluding Remarks.* According to the considered scheme it is possible to analyze some other analogous problems of delayed fracture for plates and shells bonded to rigid (or elastic) bodies. The constants λ and n can be found from special tests (bending tests, "blister" tests, etc.).

Chapter 3

CREEP AND FRACTURE UNDER MULTIAXIAL STRESS

3.1 Creep Equations

(a) *Theory of Flow.* Neglecting elastic deformation and considering only the steady creep, leads to a simple variant of the theory of creep. The metal is incompressible, i.e.

$$\dot{\varepsilon}_{ij}\delta_{ij}=0\ ,\ (i,j=1,2,3)$$

where $\dot{\varepsilon}_{ij}$ are strain rate components, and δ_{ij} is the Kronecker delta.

We use Einstein's summation convention (summation over an index which is repeated once).

We assume that the principal axes of the stress tensor σ_{ij} and strain rate tensor $\dot{\varepsilon}_{ij}$ coincide.

Let the corresponding Mohr diagrams be similar. Then,

$$\dot{\varepsilon}_{ij}=\lambda s_{ij}\ , \tag{3.1}$$

where s_{ij} are components of the stress deviator and λ is a scalar multiplier. Contracting (3.1), we obtain

$$H=2\lambda T\ , \tag{3.2}$$

where

$$H=(2\dot{\varepsilon}_{ij}\dot{\varepsilon}_{ij})^{1/2}$$

is the intensity of shear strain rates, and

$$T=(\frac{1}{2}s_{ij}s_{ij})^{1/2}$$

is the intensity of the shear stresses. In this theory, it is assumed that the creep does not depend on the hydrostatic pressure $\sigma=1/3\sigma_{ij}\delta_{ij}$.

The intensity H is a given function of the intensity T, i.e.,

$$H=f(T)T \tag{3.3}$$

or, inversely,

$$T=g(H)H .$$

Thus,

$$\lambda = \frac{H}{2T}=\frac{1}{2}f(T)=\frac{1}{2g(H)} .$$

Hence,

$$2\dot{\varepsilon}_{ij}=f(T)s_{ij} \text{ or } s_{ij}=2g(H)\dot{\varepsilon}_{ij} . \tag{3.4}$$

In the case of the power law

$$H=BT^m \text{ or } T=\bar{B}H^{\mu} , \tag{3.5}$$

where $B>0$ and $m\geq 1$ are material constants,

$$\mu=\frac{1}{m} , \bar{B}=B^{-\mu} , 0<\mu\leq 1 .$$

The coefficient B is related to the previous coefficient B_1 (for uniaxial stress, Chapter 2) as follows:

$$B=3^{m+1/2}B_1 .$$

Note that

$$f(T)=BT^{m-1} , g(H)=\bar{B}H^{\mu-1} . \tag{3.6}$$

In the linear case, $m=1$, we obtain the equations of a Newtonian viscous fluid. It is not difficult to see that

$$s_{ij}=\frac{\partial T^2}{\partial\sigma_{ij}}=\frac{\partial T^2}{\partial s_{ij}} . \tag{3.7}$$

Hence, the creep equations (3.5) can be rewritten in the form

$$2\dot{\varepsilon}_{ij}=f(T)\frac{\partial T^2}{\partial\sigma_{ij}} . \tag{3.8}$$

So, in the stress space, the vector $\dot{\varepsilon}_{ij}$ is normal to the surface (*flow surface*) $T^2=const.$

Introducing a new convex flow surface

$$F(\sigma_{ij})=const. \ ,$$

and using the same law of normality, we obtain a more general variant of the theory of flow, i.e.,

$$2\dot{\varepsilon}_{ij}=\lambda\frac{\partial F}{\partial \sigma_{ij}} \ ,$$

where λ is a scalar multiplier.

Let us introduce the potential

$$L(\dot{\varepsilon}_{ij})=\int_o^H TdH=\int_o^H g(H)HdH \ . \tag{3.9}$$

Then, the creep equations can be represented in the form

$$s_{ij}=\frac{\partial L}{\partial \dot{\varepsilon}_{ij}} \ . \tag{3.10}$$

In the case of a power law,

$$L=\frac{\overline{B}}{\mu+1}H^{\mu+1} \ . \tag{3.11}$$

If the differences of the temperatures θ for a non-uniformly heated body are not too large, the index m can be considered as a constant within the given interval of the temperatures, and the coefficient B as a function of the temperature, i.e.

$$B=B(\theta) \ .$$

(b) *Hardening Theory.* The hardening theory describes creep more comprehensively. In this theory,

$$\lambda=\lambda(T,\Gamma) \ , \tag{3.12}$$

where the Odquist's parameter Γ, characterizing the accumulated creep strain, is

$$\Gamma=\int_o^t Hdt \ .$$

(c) *Creep-Plasic Body.* In this case, the total strain rates consist of the creep strain rates $\dot{\varepsilon}_{ij}^c$ and of the instantaneous plastic strain rates $\dot{\varepsilon}_{ij}^p$, i.e.

$$\dot{\varepsilon}_{ij}=\dot{\varepsilon}_{ij}^c+\dot{\varepsilon}_{ij}^p \ . \tag{3.13}$$

The creep strain rates are described by the theory of flow or by the hardening theory. In the case of ideal plasticity, the plastic strain rates are given by Mises' relations

$$\dot{\varepsilon}_{ij}^{p}=\lambda' s_{ij}\ , \qquad (3.14)$$

where $\lambda' \geq 0$ is an arbitrary scalar multiplier.

Mises' condition of plasticity

$$T^2 = const. = \tau_y^2 \qquad (3.15)$$

should be added to the relation (3.14); τ_y is the yield stress in the case of shear.

3.2 Criteria of Fracture. Kinetic Equation of Damage

(a) *Effective Stress.* Time to brittle fracture t' under uniaxial tension is given by the formula (2.27)

$$t' = [(n+1)A\sigma_o^n]^{-1}\ ,$$

where σ_o is the tensile stress.

Under multiaxial stress it is convenient to introduce a certain effective stress $\bar{\sigma}$ in order to express the time to brittle fracture in a form similar to the uniaxial case:

$$t' = f(\bar{\sigma})\ . \qquad (3.16)$$

The simplest assumption

$$\bar{\sigma} = \sigma_1\ ,$$

where σ_1 is the maximum tensile principal stress, agrees with the experimental data for certain metals (for copper and some steels, for instance). A somewhat more complicated criterion, i.e.

$$\bar{\sigma} = \alpha\sigma_1 + (1-\alpha)T\ ,$$

where $0 \leq \alpha \leq 1$ is some constant, yields results that are in good agreement with experimental data.

A more general expression, namely

$$\bar{\sigma} = \alpha\sigma_1 + \beta T + \gamma\sigma\ ,$$

where α, β, γ are constants, was suggested by Hayhurst [16].

It is useful to note (especially for the case of the plane stress state that the intensity T can be approximately (with an accuracy of up to 7%) replaced by the maximum shear stress τ_{max}.

(b) *Kinetic Equation of Damage.* We assume that the rate of damage growth is determined by the actual effective stress $\bar{\sigma}/\psi$, i.e.,

$$\frac{d\psi}{dt}=f\left(\frac{\bar{\sigma}}{\psi}\right).$$

Consider, for simplicity, the case of a power law:

$$\frac{d\psi}{dt}=-A\left(\frac{\bar{\sigma}}{\psi}\right)^n, \tag{3.17}$$

where $A>0$ and $n\geq 1$ are material constants.

The power law can be considered as a reasonable approximation.

3.3 Front of Fracture

If the stress state is non-uniform, two stages of fracture must be considered. At the first stage (stage of latent fracture) $0\leq t\ \leq t_I$, the continuity ψ is positive at each point of the body. At the moment t_I fracture occurs at a certain point (or region) of the body. The continuous damage accumulation becomes unstable and macrocracks form. A rigorous analysis of nucleation and the development of casual macrocracks is practically impossible here. However, the same scheme of diffused damage can be used for the final stage of fracture $t>t_I$ if we introduce the concept of the moving front of fracture.

Thus, let fracture, at a moment $t>t_I$, be spread over the region V_2, Figure 3.1. The region V_2 is separated from the rest of the body V_1 (where $\psi>0$) by the moving surface Σ (front of fracture). On the surface Σ, $\psi=0$. Therefore, the total derivative is

$$\frac{d\psi}{dt}=\frac{\partial\psi}{\partial t}+\frac{\partial\psi}{\partial u}\frac{du}{dt}=0, \tag{3.18}$$

where u is the distance in the direction of the propagation of the front.

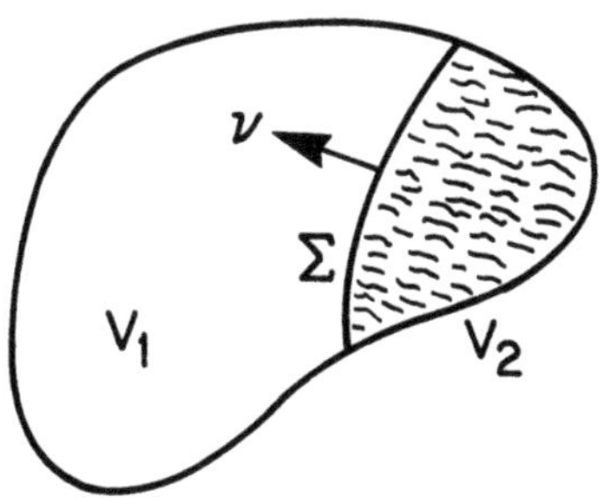

Figure 3.1

Using (3.17) and the relation for ψ analogously to (2.51), we obtain the equation of motion of the fracture front as

$$\frac{du}{dt}=(\bar{\sigma})^n_\Sigma[\frac{\partial}{\partial u}\int_o^t\bar{\sigma}^n\,dt]^{-1}_\Sigma\,. \tag{3.19}$$

The index Σ indicates that the mentioned values are calculated on the front. Note that $t \geq t_I$.

The equation of the fracture front may be written in a different form which is sometimes more convenient. The effective stress $\bar{\sigma}$ at a certain fixed point of the body at a moment τ is a function of τ and of the coordinates of the point. At the moment t, the fracture front reaches this point. Therefore $\psi=0$, and, hence,

$$A(n+1)\int_o^t\bar{\sigma}^n(\tau)d\tau=1\,. \tag{3.20}$$

Equation (3.19) can be easily derived from equation (3.20).

3.4 Fracture Time for a Round Shaft in Torsion

(a) *The Stage of Latent Fracture.* Consider the problem of brittle fracture of a circular shaft (radius a_o) twisted by a constant moment M.

The distribution of the shear stress $\tau_{\phi z}$ in the state of steady creep (see, for example [9]) is given by

$$\tau_{\phi z}=\frac{(3-\mu)M}{2\pi a_o^3}(\frac{r}{a_o})^{\mu}\,, \tag{3.21}$$

where r is the current radius. Note that we can take $\bar{\sigma}=\tau_{\phi z}$. The shear stress $\tau_{\phi z}$ reaches the maximum value for $r=a_o$.

At the moment $t=t_I$, the outer layer of the shaft fractures (i.e. $\psi=0$). Similarly to the relation (2.27) we find that

$$t_I=[(n+1)A\,\tau^n_{\phi z}(a_o)]^{-1}\,. \tag{3.22}$$

(b) *Propagation of the Fracture Front.* At the second stage $t>t_I$, the diameter of the non-fractured core of the cross-section is $2a<2a_o$ and decreases with time.

Distribution of the shear stress $\tau_{\phi z}$ in a round shaft of radius a is described by the same formula (3.21) with a_o substituted by a.

Integration of equation (3.17) for a certain value r (which, at the moment of time t, belongs to the core $r<a$) yields

$$\psi^{n+1}-\psi_I^{n+1}=-(n+1)A[\frac{(3+\mu)M}{2\pi}]^n r^{n/m}\int_{t_I}^t a^{-3n-n/m}dt\;, \tag{3.23}$$

where ψ_I denotes the distribution of ψ at the moment t_I, i.e.

$$\psi_I = 1-(n+1)A(\tau^o_{\phi z})^n t_I \, . \tag{3.24}$$

Differentiating equation (3.23) with respect to r, eliminating the derivative $\partial\psi_I/\partial r$ according to (3.24), and assuming $r=a$, we find the value of the derivative $\partial\psi/\partial r$ on the fracture front $r=a$. Furthermore, making use of the differential equation (3.17), we obtain

$$(\frac{\partial\psi}{\partial t})_{r=a} = -A(\frac{\tau^o_{\phi z}}{\psi})^n_{r=a} \, .$$

Now, according to the condition (3.18), the velocity of the motion of the fracture front is

$$\frac{da}{dt} = -\frac{m}{n}a^{-s+1}[a^{-s}t_I + \int_{t_I}^{t} a^{-s}dt]^{-1} \, , \tag{3.25}$$

where the notation

$$3n+\frac{n}{m}=s$$

is introduced. Denoting $a^s=R$, differentiating with respect to time and eliminating the integral term, we obtain the differential equation

$$\frac{d}{dt}(\ell n\frac{dR}{dt} - \frac{1}{3m+1}\ell nR)=0 \, . \tag{3.26}$$

The initial conditions are

$$a=a_o \, , \quad \frac{dR}{dt}=\frac{3m+1}{t_I}a^s_o \text{ at } t=t_I \, .$$

The value (dR/dt) is found from (3.25).

The differential equation (3.26) has a simple solution, i.e.,

$$R^{3m/3m+1}=C_1t+C_2 \, ,$$

where C_1 and C_2 are constants of integration. Determining them from the initial conditions, and assuming $R=0$, we find the time to fracture as

$$t'=t_I(1+\frac{1}{3m}) \, .$$

Since, generally, m exceeds 1 considerably, the second stage in the considered problem constitutes a small fraction of the first stage. Note that in other problems of brittle fracture the second stage can be rather long.

3.5 Fracture of Tubes under Internal Pressure

(a) *Thin-Walled Tube.* Consider, first, the simple problem of brittle fracture of a thin-walled tube under the internal pressure p. In this case, Figure 3.2, the ratio

$$\frac{b_o - a_o}{2(b_o + a_o)} << 1 ,$$

where a_o, b_o are the tube radii in the initial state. The tube has bottoms. Therefore, the tube is under an internal pressure force operating along the tube's axis. Then, the stresses in the tube are, approximately,

$$\sigma_r \approx 0 , \sigma_\phi = \frac{pa_o}{b_o - a_o} , \sigma_z = \frac{1}{2}\sigma_\phi .$$

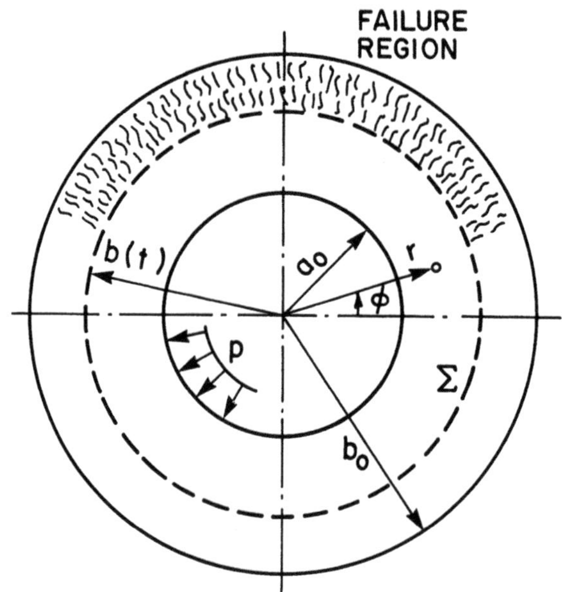

Figure 3.2

It is evident that the maximum tensile stress is

$$\sigma_1 = \sigma_\phi .$$

The corresponding time to fracture is

$$t' = [(n+1)A\sigma_\phi^n]^{-1} . \tag{3.27}$$

(b) *Thick-Walled Tube.* Stresses σ_r, σ_ϕ, σ_z (in cylindrical coordinates r, ϕ, z) are functions of r. The stress field in a tube of the dimensions a_o, b, in the state of steady creep is given by the well-known formulae

$$\sigma_r = s[1-(\frac{b}{r})^{2\mu}] \ , \ \sigma_\phi = s[1+(2\mu-1)(\frac{b}{r})^{2\mu}] \ , \qquad (3.28)$$

where $s=s(\beta)=p(\beta^{2\mu}-1)$ and $\beta=b/a_o$, $\beta_o=b_o/a_o$, $s_o=s(\beta_o)$.

If the creep index $m=2$, the stress σ_ϕ is constant and fracture occurs at all points of the tube simultaneously. If $m<2$, fracture spreads from the inner surface of the tube. If $m>2$ and $\beta \leq (1-2\mu)^{-m/2}$, the stress σ_ϕ is tensile everywhere, and reaches maximum on the outer surface of the tube. In this case, fracture develops according to the scheme shown in Figure 3.2. Since mostly, $m>2$, only this case is analyzed below.

The time of latent fracture is

$$t_l = [(n+1)A(2\mu s)^n]^{-1} \ .$$

At the time $t=t_l$ a fracture front is formed at the outer surface of the tube. Let $b(t)$ be the radius of the front at the moment t. We derive here the equation of the motion of the front. Taking $\bar{\sigma}=\sigma_\phi$ in the equation (3.19), assuming $s=s[\beta(\tau)]$, $b=b(\tau)$, $r/a_o=\beta(\tau)$, and differentiating the obtained relation with respect to t, we obtain an integro-differential equation for $\beta(t)$, i.e.,

$$\frac{d\beta}{dt} = -\frac{s^n(\beta)}{\Phi(\beta,\mu,n)} \ , \qquad (3.29)$$

where

$$\Phi(\beta,\mu,n) = \frac{n(1-2\mu)}{(2\mu)^{n-1}} \int_o^t s^n[\beta(\tau)][1+(2\mu-1)(\frac{\beta(\tau)}{\beta(t)})^{2\mu}]^{n-1} (\frac{\beta(\tau)}{\beta(t)})^{2\mu} \frac{d\tau}{\beta(\tau)} \ .$$

The initial condition is

$$\beta=\beta_o \ \text{at} \ t=t_l \ .$$

Equation (3.29) can be solved numerically by using the Euler's method, for example. It is expedient to break the integral into a sum of two integrals: over the intervals $0 \leq \tau \leq t_l$ and $t_l < \tau \leq t$. Since $\beta(\tau)=const=\beta_o$ in the first interval, the first integral is easily calculated. Figure 3.3 shows the functions $\beta=\beta(t/t_l)$ for the cases $\beta=2$, $m=4$ and $m=6$. It is assumed that $n=m$.

Note that the front propagation starts slowly. But with time its velocity increases. The duration of fracture propagation can constitute a considerable part of the stage of latent fracture. In other words, the tube with a partially fractured cross-section can resist the internal pressure during a relatively long period of time.

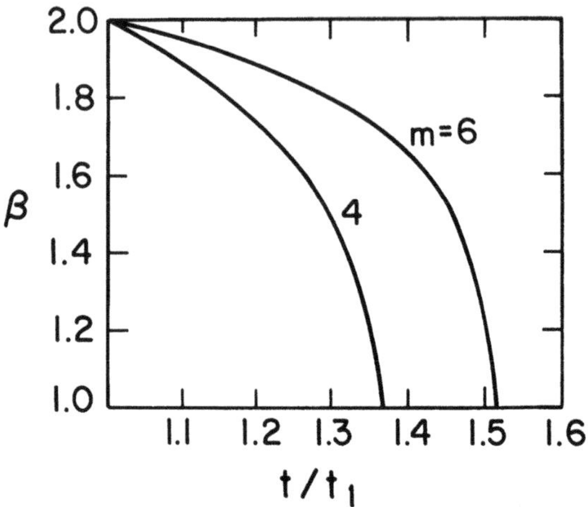

Figure 3.3

In the process of brittle fracture of tubes under internal pressure, the outer surface of tubes is often covered with a net of cracks; this is associated, to some extent, with the development of the fracture front. Casually, one of the cracks can propagate through the wall, Figure 3.4, thus bringing the normal usage of the tube to an end. An approximate analysis of this process will be considered in Chapter 4.

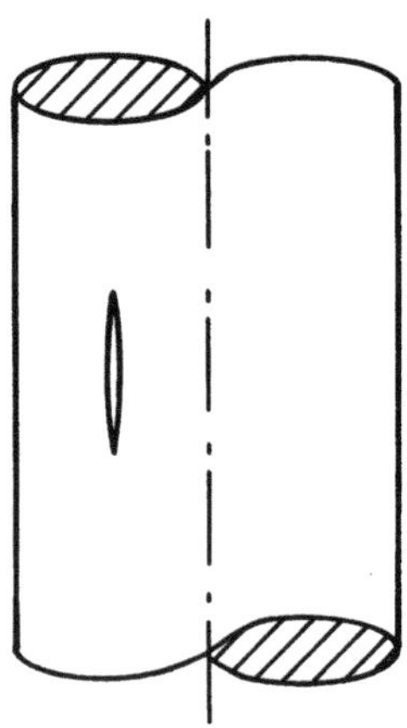

Figure 3.4

(c) *Brittle viscous Fracture of a Thin-Walled Tube.* Consider the brittle fracture of a tube subjected to large strain. The rate of continuity decrease is given by

$$\frac{d\psi}{dt} = -A\left(\frac{pa}{h\psi}\right)^n . \tag{3.30}$$

If the strain is small, $pa/h = const.$, and the time to brittle fracture can be found easily.

In the general case, when damage accumulation is accompanied by considerable creep strain, the current value of (pa/h) should be substituted into (3.30). For a thin-walled tube $h/a << 1$, and the intensity of shear stresses is

$$T = \frac{pa}{2h} .$$

The condition of incompressibility can be written in the form

$$ah = a_o h_o ,$$

where a_o, h_o, are the initial radius and thickness of the tube, respectively.

According to the equations of steady creep, the strain rates are

$$\dot{\varepsilon}_\phi = -\dot{\varepsilon}_r = \frac{1}{2} B\left(\frac{pa}{2h}\right)^m , \ \dot{\varepsilon}_z = 0 . \tag{3.31}$$

The strain rate $\dot{\varepsilon}_\phi$ is defined as

$$\dot{\varepsilon}_\phi = \frac{1}{a}\frac{da}{dt} .$$

Eliminating h, we obtain

$$\left(\frac{a}{a_o}\right)^{-1-2m} d\left(\frac{a}{a_o}\right) = \dot{\varepsilon}_{\phi o} dt = ,$$

where

$$\dot{\varepsilon}_{\phi o} = \frac{1}{2} B\left(\frac{pa_o}{2h_o}\right)^m$$

is the initial strain rate. Integrating the differential equation with the initial condition

$$a = a_o , \ \text{at } t = 0 ,$$

we obtain

$$1 - \left(\frac{a_o}{a}\right)^{2m} = 2m\dot{\varepsilon}_{\phi o} t .$$

Assuming $a \to \infty$, we find the time to viscous fracture as

$$t_1 = \frac{1}{2m\dot{\varepsilon}_{\phi o}} . \tag{3.32}$$

Introducing pa/h into the kinetic equation (3.30) according to the above considered solution, we obtain

$$(n+1)\psi^n d\psi = -\frac{1}{t'}(1-\frac{t}{t_1})^{-n/m} dt ,$$

where t' is defined by (3.27). The latter equation is similar to the equation used in the case of uniaxial tension (section. 2.4) and is valid for $t \leq t_1$.

Performing the integration, we find that the time to brittle fracture, t_*, corresponding to $\psi = 0$ in the conditions of increasing creep strain, is

$$\frac{t_*}{t_1} = 1-(1-\frac{m-n}{m}\frac{t'}{t_1})^{m/m-n} , \; m \neq n .$$

This formula coincides with the formula (2.44) for a rod under tension. But, t_1 and t' have different values. The solution is valid for $t_* \leq t_1$. Therefore,

$$\sigma_{\phi o}^{m-1} \leq 2^m \frac{n+1}{m-n}\frac{A}{B} .$$

At higher stress, viscous fracture occurs. The graph in the plane $\log \sigma_{\phi o}$, $\log t_*$ is analogous to the graph for the case of simple tension, Figure 2.13.

(d) *Brittle Fracture of a Thick-Walled Tube under Finite Deformation.* The solution of the problem of finite creep deformation of a uniformly heated tube is well-known (see, for example, [9,10]). Let $a(t)$, $b(t)$ be the current and a, b_o, the initial radii of the tube. Let

$$\beta = \frac{b}{a} , \; \beta_o = \frac{b_o}{a_o} .$$

The condition of incompressibility has the form

$$(\frac{a}{a_o})^2 = \frac{\beta_o^2 - 1}{\beta^2 - 1} .$$

If the initial position of a certain particle is characterized by the radius r_o, its position at a moment t will be at a distance r, while

$$r^2 - a^2 = r_o^2 - a_o^2 . \tag{3.33}$$

Introducing $a(t)$ here, we obtain r as a function of r_o and t.

The steady creep state in the tube (radii a_o, b) is given by the formulae (3.28). Note that

$$T=\mu s\left(\frac{b}{r}\right)^{2\mu} .$$

We also have

$$\dot{\varepsilon}_z=0 \ , \ \dot{\varepsilon}_\phi=-\dot{\varepsilon}_r \ ,$$

$$H=2\dot{\varepsilon}_\phi \ , \ \dot{\varepsilon}_\phi=\frac{v}{r} \ , \tag{3.34}$$

where v is the radial velocity.

Thus, according to the creep law (3.5)

$$\frac{v}{r}=\frac{1}{2}BT^m . \tag{3.35}$$

Note that

$$v\,|_{r=a}=\frac{da}{dt} \ , \ v\,|_{r=b}=\frac{db}{dt} .$$

It is easy to see that

$$\frac{d\beta}{dt}=\frac{1}{a}\left(\frac{d\beta}{dt}-\frac{b}{a}\frac{da}{dt}\right) .$$

Introducing here da/dt, db/dt according to (3.35), we obtain

$$(1-\beta^{-2/m})^m\frac{\beta d\beta}{\beta^2-1}=-Cp^m dt \ ,$$

$$C=\frac{1}{2}B\mu^m .$$

Let us introduce the function

$$\Phi(\beta)=\int_1^\beta(1-\beta^{-2/m})^m\frac{\beta d\beta}{\beta^2-1} \ , \ \Phi'(\beta)>0 .$$

Then, the solution of the problem of finite creep deformation of the tube is

$$Cp^m t=\Phi(\beta_o)-\Phi(\beta) . \tag{3.36}$$

The stress $\sigma_\phi(r_o,t)$ at the distance r is

$$\sigma_\phi=s\left[1+(2\mu-1)\left(\frac{b}{r}\right)^{2\mu}\right] .$$

According to the kinetic equation,

$$\psi^{n+1}=1-(n+1)A\int_o^t\sigma_\phi^n(r_o,\tau)d\tau .$$

For $m>2$, the stress is maximal on the outer surface of the tube $b(t)$; the surface will fracture ($\psi=0$) at the moment t_I, i.e.

$$\int_o^{t_1} s^n(\beta)d\tau=[(n+1)A(2\mu)^n]^{-1}.$$

Consider the stage of fracture $t>t_I$. The rate of increase of the inner radius is

$$\frac{da}{dt}=\frac{1}{2}B\mu^m as^m(\frac{b}{a})^2. \tag{3.37}$$

The rate of increase of the outer radius of the tube is determined by the creep of the tube and the propagation of the fracture front, i.e.

$$\frac{db}{dt}=(\frac{db}{dt})_c+(\frac{db}{dt})_f,$$

where

$$(\frac{db}{dt})_c=b(\dot{\varepsilon}_\phi)_{r=b}=\frac{1}{2}B(\mu s)^m. \tag{3.38}$$

On the fracture front we have

$$\int_o^t s^n[\beta(\tau)]\{1+(2\mu-1)[\frac{\beta(\tau)}{r}]^{2\mu}\}^n d\tau=[(n+1)A]^{-1}, \tag{3.39}$$

where r is the current position of the particle with the initial position $r=r_o$ (at $t=0$), and belonging to the fracture front $b(t)$ at the moment t. Using the relation (3.33) at the moments τ and t, we obtain

$$r^2=b^2(t)-a^2(t)+a^2(\tau).$$

Differentiating (3.39) with respect to time, substituting da/dt from (3.37), and making use of (3.38), we find

$$\frac{db}{dt}=\frac{1}{2}Bb(\mu s)^m-\frac{s^n[\beta(t)]}{b\Phi_1(t,\mu,n)}, \tag{3.40}$$

where

$$\Phi_1(t,\mu,n)=n(1-2\mu)(\frac{m}{2})^{n-1}\int_o^t s^n[\beta(\tau)]\{1+$$

$$+(2\mu-1)[\frac{b(\tau)}{r}]^{2\mu}\}^{n-1}\frac{[b(\tau)]^{2\mu}}{r^{2+2\mu}}d\tau.$$

Thus, the problem is reduced to the integration of the system of two equations (3.40) and (3.37) with the initial conditions

$$a=a_I\ ,\ b=b_I\ \text{at}\ t=t_I\ ,$$

where a_I, b_I are the values of a and b at the moment t_I determined according to the solution of the creep problem for the tube.

The solution is obtained numerically up to the moment t_* for which either $a=b$ (brittle fracture), or a and b increase indefinitely (viscous fracture). If there is no damage, $A=0$; if creep can be neglected $B=0$, $r=b(t)$ and the scheme, considered in section (b), can be used. Of special practical interest is the case when the strains are relatively small (say, up to 5%). But there can be a significant increase of stresses (for instance, for $\beta=1.5$ and $m=6$ there is a 14% increase of maximal stress). The analysis can be simplified in the case of relatively small deformation.

(e) *Brittle Fracture of a Non-Uniformly Heated Thick-Walled Tube.* Let the temperature in the tube be stationary, axisymmetric and independent of the z-coordinate. The temperatures $\theta(a)$ and $\theta(b)$ of the inner and outer surfaces are given.

The formulae for the stresses in a non-uniformly heated tube are the same as in the case of a uniformly heated tube, provided (see [9]) the index μ is substituted by

$$\mu_*=\mu(1+\frac{1}{2}c\theta_*) ,$$

where

$$\theta_*=\frac{\theta(b)-\theta(a)}{\ell nb-\ell na} ,$$

and c is the constant in the relation $B(\theta)=B_o e^{c\theta}$. The multiplier s will have the corresponding value s_*. Under small strain, the scheme given in section 2.8 can be used.

It must be assumed that in the formula (3.28) for σ_ϕ, $a=a_o$, $b=b(t)$ if $m_*>2$.

The fracture front propagates from the outer boundary to the inner one.

The temperature field will be stationary if the fractured layer is assumed to preserve its initial heat conductivity.

The time to brittle fracture is determined by the solution of the integro-differential equation

$$\frac{d\beta}{dt}=-\frac{s_*^n(\beta)}{\Phi(\beta,\mu_*,n)+\Psi(\beta,\mu_*,n)}$$

with the initial condition

$$\beta=\beta_o \text{ at } t=t_I ,$$

where t_1 is the time of latent fracture.

The function $\Psi(\beta,\mu_*,n)$ is the same as in the case of a uniformly heated tube, with μ being substituted by μ_*.

The function $\Psi(\beta,\mu_*,n)$ is given by

$$\Psi(\beta,\mu_*,n) = \frac{\alpha\theta_*}{2\mu_* n(1-2\mu_*)\beta}\int_o^t s_*^n[\beta(\tau)]\{1+$$

$$+ (2\mu_* - 1)[\frac{\beta(\tau)}{\beta(t)}]^{2\mu_*}\}^n d\tau .$$

For a uniformly heated tube, $\Psi(\beta,\mu_*,n)=0$. The solution of the integro-differential equation is to be found numerically.

(f) *Concluding Remarks.*

1. In the problem of brittle fracture of tubes, considered above, the stress field was assumed to be a steady one. However, the stage of steady creep begins after a period of stress redistribution; and this period of redistribution generally influences the time to brittle fracture. An approximate method describing this influence is given in [10].

2. The problem of the fracture of a thick-walled tube was also considered by Chrzanowski [20]. But, he used the hardening theory of creep as the basis.

3.6 Brittle Fracture of a Disk with a Hole

(a) *Formulation of the Problem.* Consider the axisymmetric problem of brittle fracture of a thin round disk of constant thickness h. Let a_o and b_o be the initial radii of the disk, Figure 3.5. The internal contour $r=a_o$ is traction-free, and the uniform tensile stress q is applied to the external contour.

We denote the radius of the front of fracture at the moment t by $a(t)$. Assume as usual that $\sigma_z=0$.

Let $v(r)$ be the radial velocity of the points of the disk; the strain rates are

$$\dot{\varepsilon}_r=\frac{dv}{dr} , \dot{\varepsilon}_\phi=\frac{v}{r} .$$

To simplify the mathematical analysis of the problem, we use the theory of maximal shear stress and the associated law of flow. Figure 3.6 shows the section $\sigma_z=0$ of the Tresca prism. The point representing the stress state (σ_r,σ_ϕ) is on one of the sides of the hexagon or in one of its vertices. According to the associated law, the flow develops in the direction normal to a face of the Tresca prism.

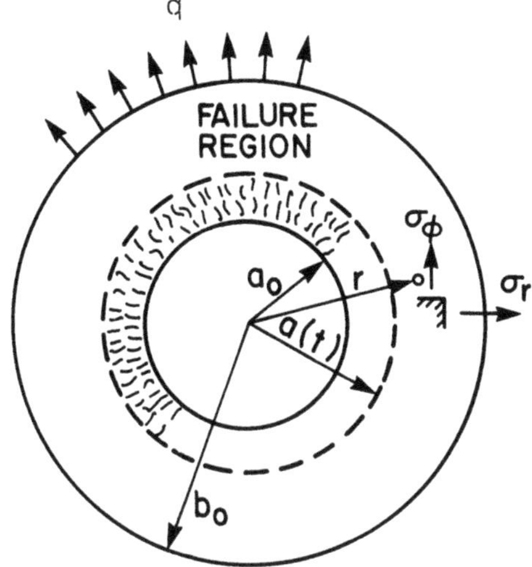

Figure 3.5

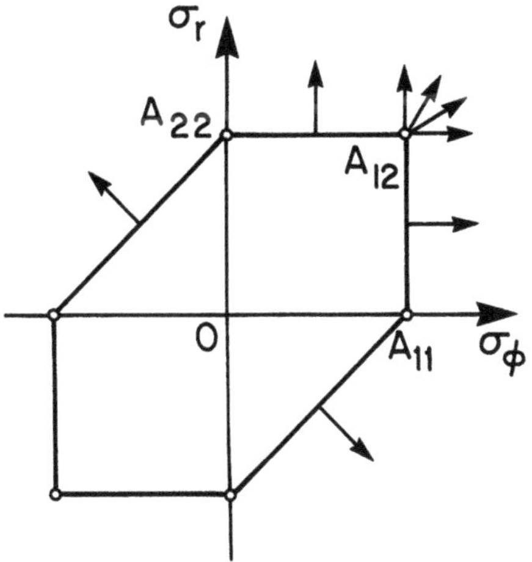

Figure 3.6

If $\sigma_\phi > \sigma_r > 0$, then $\sigma_{max} = 1/2\sigma_\phi$, and the point representing the stress state is within the line A_{11}, A_{12}; the normal is perpendicular to the σ_r-axis. Therefore, in the case of a power creep law, we have

$$\dot{\varepsilon}_r=0\ ,\ \dot{\varepsilon}_\phi=B_1\sigma_\phi^m\ ,\ \dot{\varepsilon}_z=-\dot{\varepsilon}_\phi\ . \tag{3.41}$$

In the vertex A_{12}, $\sigma_r=\sigma_\phi$, and the flow can develop in any direction within the angle formed by the normals to the adjacent sides.

Generally, there can be several zones in the disk with different regimes of flow.

We consider here only the case when the flow (3.41) takes place in the entire disk. As will be seen below, in this case, the ratio b_o/a_o is to be limited.

The compatibility condition for strain rates, and the equation of equilibrium are

$$\frac{d\dot{\varepsilon}_\phi}{dr}+\frac{\dot{\varepsilon}_\phi-\dot{\varepsilon}_r}{r}=0\ ,$$

$$\frac{d}{dr}(r\sigma_r)-\sigma_\phi=0\ .$$

The boundary conditions are

$$\sigma_r=0 \text{ at } r=a_o\ ,$$

$$\sigma_r=q \text{ at } r=b_o\ .$$

(b) *Integration.* If the flow (3.41) takes place in the disk, $dv/dr=0$. Therefore,

$$v=v_o(t)\ ,\ \dot{\varepsilon}_\phi=\frac{v_o(t)}{r}\ ,$$

where $v_o(t)$ is an arbitrary function of time. Calculating σ_ϕ from (3.41), introducing it into the equation of equilibrium, performing integration, and using the boundary conditions, we obtain

$$r\sigma_r=b_oq\frac{r^{1-\mu}-a_o^{1-\mu}}{b_o^{1-\mu}-a_o^{1-\mu}}\ ,$$

$$\sigma_\phi=b_oq\frac{r^{-\mu}}{b_o^{1-\mu}-a_o^{1-\mu}}(1-\mu)\ .$$

The solution is valid if $0<\sigma_r/\sigma_\phi<1$. It implies that

$$b_o<a_om^{m/m-1}\ .$$

In terms of dimensionless variables

$$\rho=\frac{r}{b_o}\ ,\ \alpha=\frac{a}{b_o}\ ,\ \alpha_o=\frac{a_o}{b_o}\ ,$$

the stress σ_ϕ is

$$\sigma_\phi = s(\alpha)\rho^{-\mu} ,$$

where it is assumed that

$$s(\alpha) = \frac{(1-\mu)q}{1-\alpha^{1-\mu}} .$$

At the stage of latent fracture $t < t_I$, we have $a = a_o$. Hence, σ_ϕ does not depend on time. From the kinetic equation of damage (3.17), (with $\bar{\sigma} = \sigma_\phi$) the time t_I is

$$t_I = a_o^{n/m}[(n+1)A_s^n(\alpha)]^{-1} .$$

At $t > t_I$, complete fracture develops in the region adjacent to the hole. Introducing $\bar{\sigma} = \sigma_\phi$ into the equation (3.20) of the propagation of the fracture front, taking $s(\alpha) = s[\alpha(t)]$, $\rho = \alpha(t)$, and differentiating with respect to time, we obtain a differential equation for $\alpha(t)$:

$$\frac{d\alpha}{dt} = \frac{m}{n}\frac{\alpha_o}{t_I}\left[\frac{s(\alpha)}{s(\alpha_o)}\right]^n\left(\frac{\alpha}{\alpha_o}\right)^{1-n/m} .$$

The initial condition is

$$\alpha = \alpha_o \text{ at } t = t_I .$$

Integration yields

$$\frac{t}{t_I} - 1 = \frac{n}{m}\frac{1}{\alpha_o}\int_{\alpha_o}^{\alpha}\left(\frac{1-\alpha^{1-\mu}}{1-\alpha_o^{1-\mu}}\right)^n\left(\frac{\alpha_o}{\alpha}\right)^{m-n/m} d\alpha . \tag{3.42}$$

Taking $\alpha = 1$, we find the time to fracture t' as being given by

$$\frac{t'}{t_I} = 1 + \frac{n}{m}\frac{1}{\alpha_o}\int_{\alpha_o}^{1}\left(\frac{1-\alpha^{1-\mu}}{1-\alpha_o^{1-\mu}}\right)^n\left(\frac{\alpha_o}{\alpha}\right)^{m-n/m} d\alpha .$$

In the case when n is an integer, the integral in (3.42) can easily be calculated. Figure 3.7 shows the development of the fracture front as time dependent for the case $m = n = 4$, $\alpha_o = 1/4$. The graph is calculated from the solution (3.42). Note that here the fracture propagation is relatively slow at initial stages. But, it accelerates and becomes catastrophic at the last stage. As $t' = 1.53\, t_I$, the disc fractures completely.

(c) *Concluding Remark.* The problem has been studied in more detail by Soderquist [19]; there, the fracture of a disk was considered under the condition of finite deformation; the equations of creep suggested by Odquist were used. Soderquist carried out experiments on long-term strength of magnesium disks at the temperature 260°C. His experimental data, in general, agree with the theoretical predictions. According to his data, the crack density decreases as the distance from the hole increases. The cracks have a clear radial orientation.

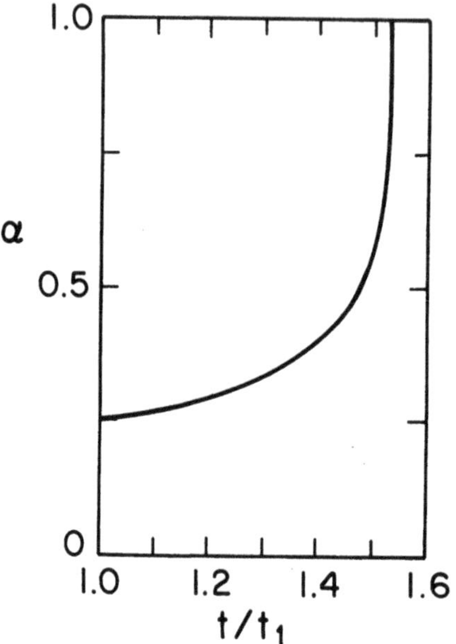

Figure 3.7

3.7 Brittle Fracture of a Thin Soft Interlayer

Consider the problem of brittle fracture of a thin soft interlayer under tension in creep conditions. Such soft (in the sense of creep resistance) interlayers can apper in metals with inhomogeneities subjected to hot working and also in welds where they appear as a result of metallurgical changes caused by high temperature. The deformation of a soft interlayer is hindered by adjacent more rigid layers, which results in high tensile stresses appearing in the interlayer.

The peculiarities of the stress state of a thin soft interlayer were first investigated by L. Prandtl in his well-known work on the flow of a thin ideally-plastic layer compressed between parallel rigid plates. His solution of the problem can be fond in textbooks on the theory of plasticity.

(a) *Steady Creep of a Thin Interlayer under Tension (Plane Strain).* Consider a plane strain problem of steady creep in a thin interlayer $|x| \leq \ell$, $|y| \leq h$, Figure 3.8, under tension. It is assumed that

$$\delta \equiv \frac{h}{\ell} \ll 1 .$$

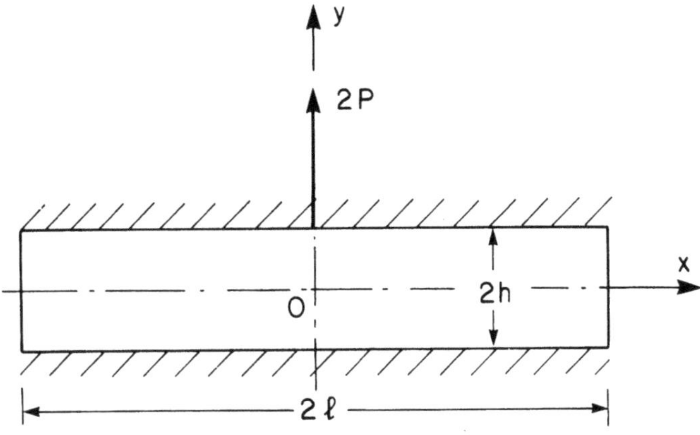

Figure 3.8

We assume that no creep deformation occurs in the rigid plates. Introduce dimensionless coordinates

$$\xi = \frac{x}{\ell} , \eta = \frac{y}{\ell} .$$

The shear stress τ_{xy} is an odd function of η. We assume that

$$\tau_{xy} = R(\xi) \frac{\eta}{\delta} , \tag{3.43}$$

whre $R(\xi)$ is an unknown function. In the case of plan strain, the stress components τ_{xz} and τ_{yz} equal zero. The velocity w (in z-direction) also equals zero. The sections $\eta=0$, $\eta=\pm\delta$ remain plane. The layer is thin. Therefore, it is natural to assume that all sections $\eta=const$ remain plane, i.e.

$$v = v(\eta) \tag{3.44}$$

where v is the velocity of the particles of the layer in the y-direction. It is obvious that $v(\eta)$ is an odd function since the middle section $\eta=0$ is a symmetry plane. The velocity component u (in x-direction), according to the adhesion condition on contact planes, is

$$u = 0 \text{ for } \eta = \pm\delta . \tag{3.45}$$

In the case of plane strain, the relations of steady creep are (see, for example [9])

$$\dot{\varepsilon}_z=0,\dot{\varepsilon}_x+\dot{\varepsilon}_y=0,\sigma_z=\frac{1}{2}(\sigma_x+\sigma_y), \tag{3.46}$$

$$\sigma_y-\sigma_x=4g(H)\dot{\varepsilon}_y,\tau_{xy}=g(H)\dot{\gamma}_{xy}. \tag{3.47}$$

where σ_x, σ_y, τ_{xy} are stress components and $\dot{\varepsilon}_x$, $\dot{\varepsilon}_y$, $\dot{\gamma}_{xy}$, are strain rate components; $g(H)$ is a characteristic function (for the given temperature) of the intensity of shear strain rates

$$H=(4\dot{\varepsilon}_x^2+\dot{\gamma}_{xy}^2)^{1/2}.$$

It is possible to assume that under creep conditions the metal is incompressible, i.e.

$$\frac{\partial u}{\partial \xi}+\frac{\partial v}{\partial \eta}=0. \tag{3.48}$$

We shall use Norton's power law. Then,

$$g(H)=\bar{B}H^{\mu-1},\ 0\leq\mu\leq 1, \tag{3.49}$$

where $\bar{B}>0$ and μ are constants. Substituting in the incompressibility equation and integrating, we obtain

$$u=-v'(\eta)\xi. \tag{3.50}$$

The arbitrary function of η can be ignored due to the symmetry condition ($u=0$ for $\xi=0$). So

$$\ell\dot{\varepsilon}_x=-v'(\eta),\ \ell\dot{\gamma}_{xy}=-v''(\eta)\xi. \tag{3.51}$$

From the adhesion condition (3.45) it follows that

$$v'=0 \text{ as } \eta=\pm\delta. \tag{3.52}$$

We seek the solution in the form

$$v''(\eta)=A\,|\eta|^{s-1}\eta, \tag{3.53}$$

where A, s are some constants. Determining $v'(\eta)$, substituting $\dot{\varepsilon}_x$, γ_{xy}, and τ_{xy} in equation (3.47) and neglecting the small term of order δ^2, we find [21] that $s=m$ and

$$R(\xi)=-\frac{A\bar{B}}{\ell}\left|\frac{A}{\ell}\right|^{\mu-1}\delta|\xi|^{\mu-1}\xi,\ (\mu=\frac{1}{m}). \tag{3.54}$$

Because of the adhesion condition (3.52), we find that along the contact plane $\dot{\varepsilon}_y=0$. Hence,

$$\sigma_y=\sigma_x \text{ at } \eta=\pm\delta. \tag{3.55}$$

The layer is thin. Therefore, it is possible to assume that the normal stress σ_y is constant in the direction normal to the layer.

Substituting the shear stress τ_{xy} into the differential equation of equilibrium,

$$\frac{\partial\sigma_x}{\partial\xi}+\frac{\partial\tau_{xy}}{\partial\eta}=0\ , \tag{3.56}$$

and integrating along the contact line, we obtain

$$\sigma_x=\frac{\bar{B}}{1+\mu}\frac{A}{\ell}\left|\frac{A}{\ell}\right|^{\mu-1}\xi^{1+\mu}+C_1\ , \tag{3.57}$$

where C_1 is an arbitrary constant. The stress σ_y, as mentioned before, can be determined by (3.57) in the whole layer. Then, the stress σ_x, inside the layer, can be found from the first equation (3.47). But, it is easy to see that the right side of the equation equals zero at $\eta=\pm\delta$ and contains the small factor δ^2. Therefore, it is possible to assume with sufficient accuracy that the stress σ_x is also constant in the y-direction.

The boundary conditions at the ends $\xi=\pm 1$ of the layer will be satisfied according to Saint-Venant's principle, i.e.

$$\int_{-\delta}^{\delta}\tau_{xy}d\eta=0\ ,\ \int_{-\delta}^{\delta}\sigma_x d\eta=0\ . \tag{3.58}$$

The first condition is satisfied. The second condition determines the constant C_1. As a result of some calculations, we obtain

$$\sigma_y=\frac{2+\mu}{1+\mu}(1-|\xi|^{1+\mu})p\ ,\ \sigma_x=\sigma_y=\sigma_z\ , \tag{3.59}$$

$$\tau_{xy}=(2+\mu)|\xi|^{\mu-1}\xi\eta p\ , \tag{3.60}$$

where $p=P/\ell$ is the mean stress. In the case of compression $p<0$.

From this solution it follows that a triaxial tension takes place in the central part of the layer.

The biggest tensile stress (at $\xi=0$) is

$$\frac{\sigma_{y,max}}{p}=1+\frac{1}{1+\mu}\ . \tag{3.61}$$

The distribution of the normal stress σ_y is shown in Figure 3.9 for $\mu=$ 0; 0.5; 1.0.

The shear stress on the contact line $\eta=\delta$ is as follows:

if $\mu=0$ then,

$$\tau_{xy}=\begin{cases}+2p\delta \text{ for } \xi>0\ ,\\ -2p\delta \text{ for } \xi<0\ ;\end{cases}$$

if $\mu=1$ (linear viscosity), τ_{xy} is proportional to ξ.

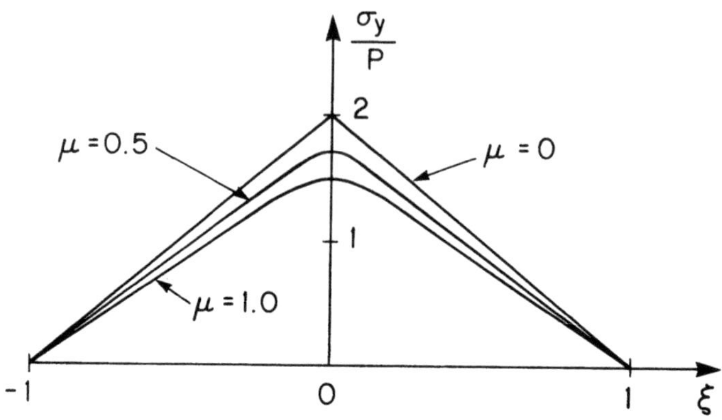

Figure 3.9

For $\mu=0$ and $p<0$ the stress distribution is analogous to the stress distribution in the above mentioned problem of Prandtl. But, in the case of creep, it occurs at any load p; in creep problems there is no yield condition and, consequently, no limit load.

(b) *Unsteady Creep.* At the initial moment $t=0$ the layer is elastic. It is possible to assume that the layer and the "rigid plates" have the same elastic constants. Then, the initial elastic stress state of the layer (we denote this state by a prime ′) is

$$\sigma'_x=0\;,\;\sigma'_y=p\;,\;\tau'_{xy}=0\,.$$

The stresses in the steady creep state considered above, will be denoted by two primes (σ''_x, σ''_y, τ''_{xy}). In unsteady creep under constant load the stress state changes monotonically from the initial elastic state to the steady creep state (see [9,21]). According to this soluton, the maximum tensile stress increases monotonically from the initial value 1 to the steady (as $t\to\infty$) value

$$\alpha=\frac{2+\mu}{1+\mu}\,.$$

Note that $1.5<\alpha\leq 2.0$.

(c) *Time to Brittle Fracture.* The time to brittle fracture of the layer is determined by the kinetic equation. Then, the time to local fracture t_* can be found from the equation

$$\frac{1}{A(n+1)}=\int_o^{t_*}\sigma_{y,max}^n(t)dt \ .$$

It is easy to see that there exists a unique root t_*. Since $p\leq\sigma_{y,max}\leq p\alpha$ we have

$$t''\leq t_*\leq t' \ ,$$

where $t'=[(n+1)Ap^n]^{-1}$ is the time to fracture for the elastic field, and $t''=t'\alpha^{-n}$ is the fracture time for a steady creep stress field.

(d) *Tension of a Thin Axisymmetric Interlayer*. Consider the problem of tension of a thin "soft" axisymmetric layer in the case of steady creep.

Let r, ϕ, z be cylindrical coordinates, u, v, w the radial, circular and axial velocity components, respectively. As before, the layer $(o\leq r<a; |z|\leq h)$ is joined to parallel rigid plates. We assume again that $\delta=h/a<<1$, and introduce the dimensionless coordinates

$$\rho=\frac{r}{a} \ , \ \zeta=\frac{z}{a} \ .$$

The case of torsion is excluded from consideration here. Therefore, $v=0$. Assuming that the sections $\zeta=const.$ remain plane, we have $w=w(\zeta)$.

The strain rate components are as follows:

$$\dot{\varepsilon}_r=\frac{1}{a}\frac{\partial u}{\partial\rho} \ , \ \dot{\varepsilon}_\phi=\frac{1}{a}\frac{u}{\rho} \ ,$$

$$\dot{\varepsilon}_z=\frac{1}{a}\frac{\partial w}{\partial\zeta} \ , \ \dot{\gamma}_{rz}=\frac{1}{a}\frac{\partial u}{\partial\zeta} \ .$$

Substituting the strain rates in the condition of incompressibility and integrating, we obtain

$$u=-\frac{1}{2}\rho w'(\zeta)+\frac{C(\zeta)}{\rho} \ , \tag{3.62}$$

where $C(\zeta)$ is an arbitrary function. The velocity u is bounded. Hence, $C(\zeta)=0$. Then,

$$\dot{\varepsilon}_r=\dot{\varepsilon}_\phi=-\frac{1}{2}\dot{\varepsilon}_z=-\frac{1}{2a}w'(\zeta) \ ,$$

$$\dot{\gamma}_{zz}=-\frac{1}{2a}\rho w''(\zeta) \ ,$$

and from the relations of steady creep it follows that $\sigma_r=\sigma_\phi$ and

$$\sigma_z-\sigma_r=\frac{3}{2}g(H)w'(\zeta) \ , \tag{3.63}$$

$$\tau_{rz} = -\frac{1}{2} g(H) \rho w''(\zeta) . \tag{3.64}$$

As before, we assume

$$\tau_{rz} = R(\rho) \frac{\zeta}{\delta} . \tag{3.65}$$

Substituting the strain rate components into the relation (3.64), using the power law, assuming that

$$w''(\zeta) = A \mid \zeta \mid^{s-1} \zeta ,$$

comparing the formula (3.65) to the formula (3.64), and neglecting a small summand, we find that $s = m$ and

$$\tau_{rz} = A_2 \rho^{\mu} \zeta , \tag{3.66}$$

where A_2 is some constant. Substituting the obtained values into (3.63), we can see that its right-hand part is small (of the order of δ^2).

Therefore, according to the order of approximation used, we have

$$\sigma_r = \sigma_\phi = \sigma_z . \tag{3.67}$$

Consider now the differential equation of equilibrium

$$\frac{\partial \sigma_r}{\partial \rho} + \frac{\sigma_r - \sigma_\phi}{\rho} + \frac{\partial \tau_{rz}}{\partial \zeta} = 0 . \tag{3.68}$$

Using (3.66) and (3.67), and integrating, we find the stress σ_r. The constant of integration is determined from the boundary condition

$$\sigma_r = 0 \text{ for } \rho = 1 .$$

The stress σ_z is assumed to be constant in the direction normal to the layer. Then, the stress components σ_r, σ_ϕ, are also constant in this direction.

Determining the constant A_2 from the condition that stresses σ_z are statically equivalent to the tensile force P, we obtain

$$\sigma_z = \frac{3+\mu}{1+\mu} p (1 - \rho^{1+\mu}) , \ \sigma_r = \sigma_\phi = \sigma_z ,$$

$$\tau_{rz} = (3+\mu) p \, \delta^{\mu} \zeta ,$$

where

$$p = \frac{P}{\pi a^2} .$$

These formulae are analogous to the formulae (3.59), (3.60) except for the factor $(3+\mu)$ used here instead of $(2+\mu)$ there. Hence, the stresses in the axisymmetric case are $(3+\mu/2+\mu)$ times higher than in the case of plane strain.

(e) *Conclusion.* In the case of plane strain considered above, the triaxial tension is realized in the central part of the layer. The maximum stress is $(2+\mu/1+\mu)$ times higher than the average stress p.

According to Prandtl's solution, the level of triaxial pressure in a thin ideally plastic layer compressed between rigid plates can be significantly higher than the average stress. It can be accounted for by the value of the shear stress on the contact planes. For an ideally plastic layer it is usually assumed that the shear stress is equal to the shear yield stress τ_y.

Here, it must be noted that the actual value of the contact shear stress is unknown. But, on the average, it is smaller than τ_y.

In the case of creep considered above, the contact shear stress is equal to $2p\delta$ (for $\mu=0$). This value following from the more realistic condition of adhesion is substantially smaller than τ_y. Combining the solutions for tension and compression, it is also possible to consider the plane strain problem of bending of a thin layer.

3.8 Constitutive Equations of Creep and Damage under Simple Loading

(a) *Creep Equations.* Constitutive equations of creep and damage under uniaxial tension were considered in section 2.4. Leckie and Hayhurst [22] extended these equations to the multiaxial stress state.

According to experiments, strain rates depend on effective stress $\bar{\sigma}$ and the components of strain rate $\dot{\varepsilon}_{ij}$ are proportional to the components s_{ij} of the stress deviator. Thus, in the case of a power law,

$$\dot{\varepsilon}_{ij}=\lambda\bar{\sigma}^{m-1}s_{ij}\ , \tag{3.69}$$

where $\lambda>0$ is a certain scalar multipler, and $m\geq 1$ is a constant index.

Comparing the form of the latter equation with the uniaxial relation (2.56) for the case $q=m$, we obtain

$$\dot{\varepsilon}_{ij}=\frac{3}{2}b\left(\frac{\bar{\sigma}}{\psi}\right)^{m-1}\frac{s_{ij}}{\psi}\ . \tag{3.70}$$

Hence,

$$\lambda=\frac{3}{2}b\frac{1}{\psi}\left(\frac{\bar{\sigma}}{\psi}\right)^{m-1}\ .$$

Note that b is the constant used previously in section 2.4.

(b) *Kinetic Equation of Damage.* Analogously to equation (3.69), when we introduced the effective stress $\bar{\sigma}$, the kinetic equation (2.57) can take the form

$$\dot{\omega}=A(\frac{\overline{\sigma}}{\psi})^n \ , \ \dot{\omega}=-\dot{\psi} \ . \tag{3.71}$$

$\omega=0$, when the material is in its undamaged state; $\omega=1$ at the moment of fracture.

It is assumed above that the metal is incompressible, the damage is isotropic, and the loading is proportional ("simple"), i.e.,

$$s_{ij}=s_{ij}^o t \ ,$$

where s_{ij}^o is a constant stress deviator and $t>0$ is an increasing parameter (such as time or loading parameter).

Using the relations (3.7), we can represent the equations (3.70) in the form

$$\dot{\varepsilon}_{ij}=\frac{3}{2}b(\frac{\overline{\sigma}}{\psi})^{m-1}\frac{1}{\psi}\frac{\partial T^2}{\partial s_{ij}} \ . \tag{3.72}$$

(c) *On Stress Redistribution.* The creep acceleration, which is described by the obtained constitutive equations, can lead to a significant stress redistribution, because in the high stress region an intensive process of creep and damage takes place.

This phenomenon is illustrated by the Hayhurst-Leckie's solution of the problem of torsion of a circular bar [23] and by their tensile tests on notched specimens [22].

It must be noted that the stress redistribution is significant for metals that are prone to brittle-viscous fracture. For the brittle fracture, the strain is small, the influence of stress concentration on time to fracture is substantial, and the stress redistribution is not so intensive.

3.9 Energy Theorem

Consider a body under steady creep conditons.

At $t<0$, the body surface S is not loaded.

At $t=0$, the surface traction T_i (i-1,2,3) is applied to the part S_T of the surface and remains constant for $t>0$. For the remaining part S_v, the specified velocity v_i^o is zero.

With time, the body will reach the steady creep state.

(a) *Basic Energy Relation.* Let σ_{ij} be some stress field that satisfies the differential equations of equilibrium in the body (for simplicity, we assume that there are no body forces), i.e.,

$$\frac{\partial \sigma_{ij}}{\partial x_j}=0\ ,\quad i,j=1,2,3$$

and is consistent with the specified traction vector T_i^o on S_T, i.e.,

$$\sigma_{ij}\nu_j=T_i^o \text{ on } S_T\ ,$$

where ν_j $(j=1,2,3)$ are components of the unit normal.

The body is incompressible, its deformation is small.

For the solid body, the basic energy relation is given by

$$\int \sigma_{ij}\dot{\varepsilon}_{ij}dV=\int T_i^o v_i dS_T\ , \tag{3.73}$$

where the first integral extends over the volume V of the body, and the second-over the surface S_T.

It is not difficult to prove this relation by making use of the Gauss' divergence theorem.

This relation implies that the power of internal forces is equal to the power of external forces.

(b) *Steady Creep in the Absence of Damage.* In this case, according to equation (3.10), we have

$$s_{ij}=\frac{\partial L}{\partial \dot{\varepsilon}_{ij}}\ ,$$

where $L(\dot{\varepsilon}_{ij})$ is a potential function.

In the case of the power law (3.5)

$$L=\frac{\overline{B}}{\mu+1}H^{\mu+1}\ ,$$

where H is the intensity of shear strain rates; $\overline{B}>0$ and $\mu=1/m$ are material constants.

The potential L is a positive homogeneous function of degree $\mu+1$.

The dissipation density rate is

$$\dot{D}=\sigma_{ij}\dot{\varepsilon}_{ij}=s_{ij}\dot{\varepsilon}_{ij}=(\mu+1)L=\overline{B}H^{\mu+1}\ . \tag{3.74}$$

Hence,

$$\int THdV=\int T_i^o v_i dS_T\ .$$

(c) *Steady Creep Accompanied by Damage.* For this case, we have the relations (3.70); contracting them, we obtain

$$H = 3b\left(\frac{\overline{\sigma}}{\psi}\right)^{m-1}\frac{T}{\psi} .$$

From the equations (3.72) we see that, when creep is accompanied by damage, we have

$$\dot{D} = TH .$$

The energy theorem can be used to obtain approximate solutions and estimates. A simple approximate method for the problems of steady creep in engineering application was proposed by Leckie and Hayhurst [22]. They introduced the notion of the representative creep stress σ_r; it is a uniaxial stress which brings about a creep dissipation rate equal to the average dissipation rate in the body.

In the case of uniaxial tension, according to the Norton's law (2.2), the creep dissipation density rate is

$$\sigma\dot{\varepsilon} = B_1\sigma^{m+1} .$$

According to definition,

$$\sigma_r = \left[\frac{1}{B_1 V}\int \dot{D}dV\right]^{1/m+1} .$$

It has been found that the representative creep stress σ_r is practically independent of the creep constants. With the representative stress known, the average creep strain of the structure can directly be estimated from the uniaxial tension test data.

Another approximate method was developed by Leckie and Hayhurst [22] for the case of a kinematically determined structure in which the velocity field has the form

$$v_i = v_{i,s}\tau(t) ,$$

where $\tau(t)$ depends solely upon the time, and $v_{i,s}$ is a steady creep velocity field that depends solely on the coordinates.

3.10 Brittle Fracture under Complex Loading

(a) *Basic Concepts.* For the case of proportional ("simple") loading, the kinetic equation considered in section 3.2 can be used. The kinetic equation for the case of cyclic loading was considered in section 2.6.

Damage accumulation under complex loading is of great interest for practical applications.

Here, the problem is that of mathematical modelling of damage and also of determining of a set of corresponding functions and constants based, as a rule, on scarce experimental data.

Some general schemes of damage description were briefly reviewed in Chapter 1.

Under non-proportional ("complex") loading, the damage cannot be considered as isotropic and cannot be represented by a scalar function.

A simple variant of the damage description will be given below. It is based on available experimental data and is applicable to engineering problems.

We introduce the damage variable as a vector function [10].

Consider an element dS in a body section, Figure 1.1; $\underset{\sim}{\nu}$ is the unit normal.

Under the action of the normal tensile stress σ_ν, the material is fractured. It is natural to consider the deterioration of the material as a system of planar microcracks perpendicular to the stress vector $\sigma_\nu \underset{\sim}{\nu}$.

The magnitude of the damage vector is defined locally by the crack density ω (analogously to the definition (1.2) in Chapter 1).

So, the damage $\underset{\sim}{\omega}=\omega_\nu \underset{\sim}{\nu}$. Note that $\omega_\nu=0$ for the undamaged section and $\omega_\nu=1$ at the moment of fracture.

We assume that damage (nucleation and growth of microcracks) develops under the action of a tensile normal stress. But, there is no damage accumulation under compression.

In general, σ_ν depends on time. We also assume that the principle of linear summation of damage is applicable here.

Sometimes, instead of ω, it is convenient to use the "continuity" vector

$$\underset{\sim}{\psi}=\psi_\nu \underset{\sim}{\nu} ,$$

where $\psi_\nu=1-\omega_\nu$.

The value ψ_ν is described by the kinetic equation of the type

$$\frac{d\psi_\nu}{dt}=\begin{cases} f(\sigma_\nu,\psi_\nu,\ldots) & \text{for } \sigma_\nu>0 , \\ 0 & \text{for } \sigma_\nu\leq 0 , \end{cases}$$

with $f(0,\psi_\nu,\ldots)=0$.

Besides, f can depend on the stress and strain invariants and on other parameters such as temperature, time (if there is an effect of aging) etc.

It is assumed that damage accumulates on the planes orthogonal to the tensile stress vector. It is convenient to use the power function

$$\frac{d}{dt}\psi_\nu = \begin{cases} -A\left(\frac{\sigma_\nu}{\psi_\nu}\right)^n & \text{for } \sigma_\nu > 0 \,, \\ 0 & \textit{for } \sigma_\nu \le 0 \,, \end{cases} \tag{3.75}$$

where A and n are constants.

Consider now the criteria for fracture. We shall distinguish between partial and complete fracture of the material.

Partial fracture can be of two types: in the one, layer stratification takes place; in the other, there is fiber disintegration.

In the first case, the material is fractured in one direction (for example - in the direction 1, i.e. $\psi_1 = 0$), but it can resist tension in the plane normal to the direction 1.

In the second case, the material is fractured in two perpendicular directions 1 and 2, but can resist tension in the direction 3, i.e. along the "fibers". Thus, the problem of strength of the material should be analyzed in compliance with the loading program.

We consider a material to be fractured partially if at the given point there is a plane on which $\psi_\nu = 0$, while $\sigma_\nu > 0$. In the case of complete fracture, $\psi_\nu = 0$ for all directions, and the material offers no resistance to tension.

When considering the planes at the given point, we should select the one with the minimal fracture time, i.e.

$$t_* = \min \,.$$

Under the conditions of a non-uniform stress field, a fracture front forms at a certain moment of time. The propagation of the front is described by the equation

$$\frac{d\psi}{dt} = \left(\frac{\partial\psi}{\partial t}\right)_\Sigma + \left(\frac{\partial\psi}{\partial u}\right)_\Sigma \frac{du}{dt} = 0 \,,$$

where u is the distance along the normal.

If the creep deformation is not small (mixed type of fracture), the damage accumulation can be characterized by equation (3.75). This equation is concerned with the considered particle of the body.

(b) *Extension of a Tube Followed by Torsion.* As an example consider the problem of brittle fracture of a thin-walled round tube under torsion proceeded by extension during the time interval $0 \le t < t_I$.

In tension, the normal stress on a plane with the normal ν forming the angle ϕ with the circumferential direction is given by

$$\sigma_v = \sigma \sin^2\phi ,$$

where σ is the tensile stress.

According to the kinetic equation, the damage at the moment t_I is

$$\psi_{vI}^{n+1} = 1 - (n+1) A \sigma_{v1}^n t_I .$$

At $t \geq t_I$, only the twisting moment is acting, and on the same plane

$$\sigma_{v2} = \tau \sin 2\phi ,$$

where τ is the torsional stress. Integrating the kinetic equation, and determining the arbitrary constant from the continuity condition

$$\psi_{v2} |_{t=t_I} = \psi_{v1} ,$$

we obtain

$$\psi_{v2}^{n+1} = 1 - (n+1) A [\sigma_{v1}^n t_I + \sigma_{v2}^n (t - t_I)] .$$

If fracture occurs (i.e. $\psi_{v2}=0$) at the moment $t=t_*$, it follows from this relation that

$$t_* - t_I = t'_o \left(1 - \frac{t_1}{t'_1} \sin^{2n}\phi\right) \sin^{-n} 2\phi ,$$

where $t'_1 = [(n+1) A \sigma^n]^{-1}$ is the time to fracture under tensile stress σ.

The time to fracture under torsion is $t'_o = [(n+1) A \tau^n]^{-1}$.

It is evident that

$$\lambda \equiv \frac{t_I}{t'_1} \leq 1 .$$

We have to find that orientation $\phi = \phi_*$ for which the fracture time is minimal.

The condition $dt_*/d\phi = 0$ yields

$$1 - 2\sin^2\phi + \lambda \sin^{2n}\phi = 0 .$$

It is evident that $\phi_* > \pi/2$ as $\lambda > 0$.

For $\lambda << 1$, fracture occurs on planes with orientations close to $\phi = \pi/2$; ϕ_* increases as λ increases. The corresponding curves are shown in Figure 3.10. Note the strong influence of non-linearity on the process of brittle fracture. For large values of n, the "dangerous" planes are close to $\phi = \pi/4$. Even for $\lambda = 1$ (when tube is almost fractured under tension), damage accumulation on these planes is relatively small. The ratio $(t_* - t_I)/t'_o$ for $\lambda = 1$ is

$$\frac{t_* - t_I}{t'_o} = \frac{1 - sin^{2n}\phi_*}{\sin^n 2\phi_*} \; .$$

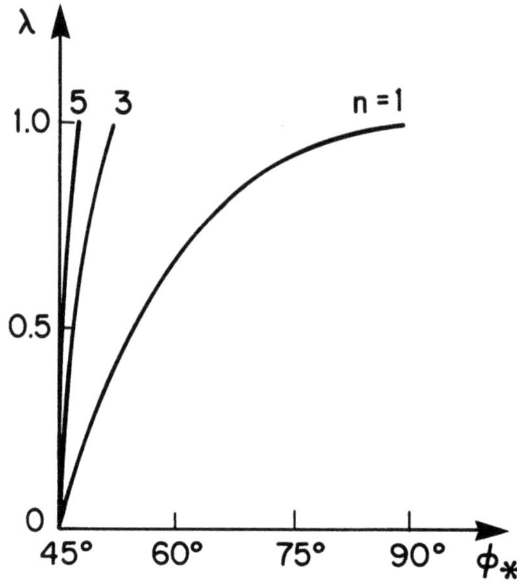

Figure 3.10

This relationship is shown in Figue 3.11. With n increasing, this ratio tends to unity, and the influence of damage accumulated under tension decreases fast ("short memory"); only for $n=1$ the tube does not resist torsion.

(c) *Successive Tensions in Different Directions.* Let the tensile stress σ' acting in the direction of the x-axis in the time interval $0 \leq t < t_I$ be followed by the tensile stress σ'' acting in the direction constituting the angle $0 \leq \alpha < \pi/2$ with the x-axis, Figure 3.12.

The normal stress on the plane with the normal ϕ is

$$\sigma_{\nu 1} = \sigma' \cos^2\phi \; , \; t < t_1 \; , \; 0 \leq \phi \leq \frac{\pi}{2} \; .$$

At the moment $t = t_1$,

$$\psi_{\nu 1}^{n+1} = 1 - \frac{t}{t'_\nu} \; ,$$

where t'_ν is the time to brittle fracture under the action of the normal stress $\sigma_{\nu 1}$, i.e.,

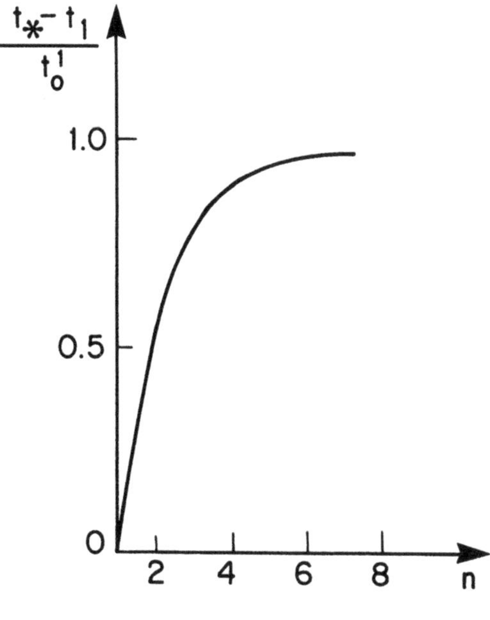

Figure 3.11

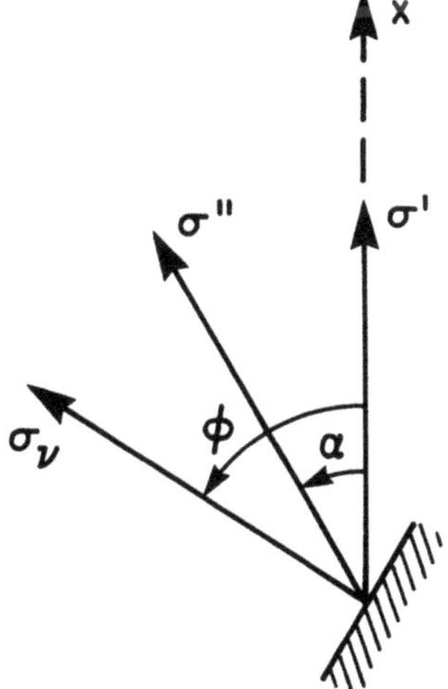

Figure 3.12

$$t'_v=[(n+1)A\sigma_{v1}^n]^{-1}.$$

At the second stage,

$$\sigma_{v2}=\sigma''\cos^2(\phi-\alpha)\ ,\ t\geq t_1\ .$$

Integrating the kinetic equation for ψ_{v2}, and determining the constant of integration from the condition

$$\psi_{v2}\,|_{t=t_1}=\psi_{v1}\ ,$$

we find that

$$\psi_{v2}^{n+1}-\psi_{v1}^{n+1}=\frac{1}{t''_v}(t_1-t)\ ,$$

where t''_v is the time to brittle fracture under the action of the normal stress σ_{v2}, i.e.,

$$t''_v=[(n+1)A\sigma_{v2}^n]^{-1}.$$

At the moment of fracture t_* we have $\psi_{v2}=0$. Hence,

$$t_*=t_1(1-\frac{t''_v}{t'_v})+t''_v\ .$$

Let us find the plane $\phi=\phi_*$ for which the time to fracture is minimal (in the case $\sigma''=\sigma'$. The condition $dt_*/d\phi=0$ yields

$$\tan\phi\,cot(\phi-\alpha)=1-(\lambda\cos^{2n}\phi)^{-1}\ ,$$

where

$$\lambda\equiv\frac{t_1}{t'}\leq 1\ ,\ t'=[(n+1)A\sigma'^n]^{-1}\ .$$

t' is the time to brittle fracture for the stress σ'.

When $\lambda<1$, the right-hand side of the obtained relation is a negative monotonically decreasing function of ϕ, Figure 3.13. The left-hand side is negative for $\phi<\alpha$ and positive for $\phi>\alpha$. Thus, there exist a unique root ϕ_* in the interval

$$0\leq\phi_*\leq\alpha\ .$$

The equality signs are realized for $\lambda=1$ and $\lambda=0$ only. It means that the normal to the plane of fracture is between the directions σ' and σ'' if fracture does not occur within the first stage (i.e. $\lambda\neq1$) or if $t_1\neq0$ (i.e. $\lambda\neq0$).

If $n>>1$, the plane of fracture is close to the plane α. This is another example of a "short memory" in the case of strong non-linearity.

(d) *Example of Successive Tensions.* Consider a simple case of successive tensions acting on the same rod. Let the tensile σ_1 act during the time interval $0\leq t\leq t_1$ be followed by the tensile stress σ_2 at $t>t_1$. Determine the time to fracture t_* according to the principle of linear summation.

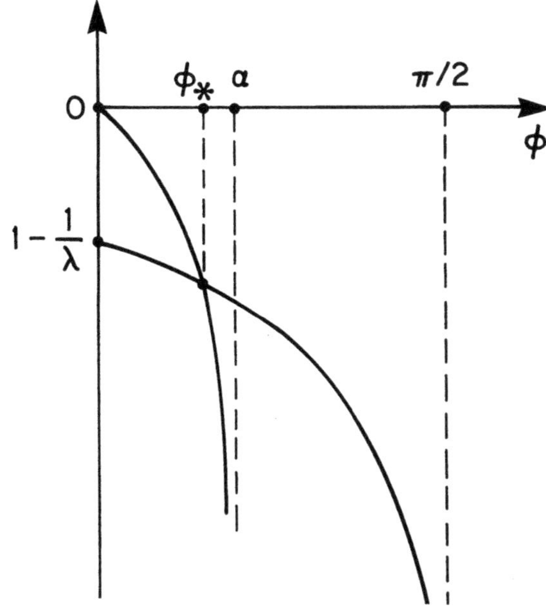

Figure 3.13

Let us assume that

$$t_2 = t_* - t_1 .$$

Then,

$$\frac{t_1}{t'_1} + \frac{t_2}{t'_2} = 1 ,$$

where

$$t'_1 = [(n+1)A\sigma_1^n]^{-1} , \; t'_2 = [(n+1)A\sigma_2^n]^{-1} .$$

It is obvious that $t_1/t'_1 < 1$. Substituting into the previous relation the value t_2, we obtain

$$\frac{t_*}{t'_2} = 1 - \frac{t_1}{t'_2}[1 - (\frac{\sigma_1}{\sigma_2})^n] .$$

Let us assume that $n >> 1$ and $\sigma_1/\sigma_2 < 1$. Then, the value in the square brackets is close to 1. If, moreover, $t_1/t'_2 << 1$, we have

$$t_* \approx t'_2 .$$

So, loading during the first time interval does not influence the fracture time in the case of strong non-linearity. If t'_1 is not small, it is necessary to add it to t'_2 i.e.,

$$t_* = t_1 + t'_2 .$$

Thus, the damage accumulation during the second time interval does not depend on that taking place during the first time interval, and the time to fracture is reckoned from the moment t_1.

(e) *The Rule of Short Memory.* Problems considered above and some other examples show that in the case of strong non-linearity ($n >> 1$), the damage that was accumulated earlier has little influence on the process at the next stage. In other words, the material does not "remember" over a long period of time the damage received at the earlier stage of loading. We assume that the behavior of the material in the case of strong non-linearity is characterized by this "rule of short memory".

(f) *Example: Torsion of a Round Shaft After Tension.* Consider the problem of brittle fracture of a round shaft undr successive tension and torsion. In the time interval $t \leq t_1$, the shaft is tensioned by the stress σ. At the time $t > t_1$, the shaft is under torsion. We can neglect a relatively short period of unsteady creep and consider the stress state of the shaft as a steady one. At $t > t_1$, the shear stress τ is given by the formula (section 3.4)

$$\tau = k\left(\frac{\tau}{a_o}\right)^{\mu} ,$$

where k is a certain constant, and a_o is the radius of the shaft. Making use of the principle of linear summation, it is not difficult to find the time to fracture of the surface layer $r = a_o$. The propagation of the fracture front at $t > t_I$ is described analogously to the case considered in section 3.4. The solution has the same form. But, instead of the time t_I, it is necessary to introduce the time t_I^o.

(g) *Tests of Leckie, Hayhurst and Trampzynski.* These authors have recently made some tests [25] on creep fracture of thin-walled copper and aluminum alloy tubes under non-proportional loading (successive tension and torsion). At the first stage $0 \leq t \leq t_1$, the tube was loaded by constant tensile stress σ and constant shear stress $\tau = 1/3\sigma$.

At the second stage $t > t_1$, the tensile stress was the same, but the tube was twisted in the opposite direction. For copper, the index $n = 5.5$; the fracture criterion is close to the maximal tensile stress σ_1. According to the rule of short memory, the fracture time is

$$t_* = t_1 + t'_2 ,$$

where t'_2 is the fracture time for the steady load test $\sigma+\tau$. This result is in good agreement with the experimental data for copper.

In the case of multi-reversal of the torsional moment at the second stage (multi-reversal torsion, constant tension tests) the life-time of the sample was twice as large as that in steady load tests. In the case of a sufficiently large number of cycles, this result is in agreement with the "dormant damage mechanism" (section 2.3). For an aluminum alloy test, results cannot be clearly interpreted. The fracture criterion is close to the intensity T of the shear stresses; the fracture process is independent of the change of the direction of the maximal tensile stress. Possibly, the fracture mechanism for the aluminum alloy is different. It stands to reason that some more tests on creep fracture under complex loading are desirable.

3.11 Creep and Damage under Complex Loading

(a) *Damage Accumulation.* Damage accumulation under complex loading is considered according to the scheme described in section 3.10. The damage accumulates on the planes orthogonal to the tensile stress vector. Under compression there is no damage accumulation.

(b) *Influence of Damage on the Creep Deformation.* Let σ_i $(i=1,2,3)$ be the principal values of the stress tensor, and x_i its principal axes.

Using the principal stresses σ_i at a certain time t, it is possible to find the corresponding values ψ_i for the principal directions basing on section 3.10.

In considering small deformations usually associated with brittle fracture, the changes in the positions of body particles can be ignored.

In the case of finite deformation, damage is associated with definite body particles. Therefore, it is expedient to use the Lagrangian description.

Let us introduce the actual stress tensor $\sigma_{i,a}$. Its principal directions coincide with the principal directions x_i; its principal values are given by

$$\sigma_{i,a}=\frac{\sigma_i}{\psi} . \tag{3.76}$$

We consider here a simple version of the constitutive equations based on the power law and on von Mises' criterion of flow. Let us denote the intensity of the actual shear stresses by

$$T_a=\frac{1}{\sqrt{6}}[(\sigma_{1,a}-\sigma_{2,a})^2+(\sigma_{2,a}-\sigma_{3,a})^2+(\sigma_{3,a}-\sigma_{1,a})^2]^{1/2} ,$$

and the principal values of the actual stress deviator by

$$s_{i,a}=\sigma_{i,a}-\sigma_a .$$

Here, $\sigma_a 1/3(\sigma_{1,a}+\sigma_{2,a}+\sigma_{3,a})$ is the corresponding hydrostatic pressure. Making use of the previous steady creep equation, it is natural to introduce the following constitutive equations:

$$\dot{\varepsilon}_i = \frac{1}{2} B T_a^m \frac{s_{i,a}}{\psi} . \tag{3.77}$$

In the case of isotropic damage, $\psi_i=\psi$, and the equations (3.77) are reduced to the constitutive equations of Leckie and Hayhurst [22].

(c) *Orthotropic Metal.* Consider the extension of the above given scheme to the case of orthotropic metal. Let the axes of orthotropy coincide with the principal axes x_i. Let ψ_i $(i=1,2,3)$ be the damage parameters in these directions. The corresponding kinetic equations have the form

$$\frac{d\psi_i}{dt} = \begin{cases} -A_i\left(\dfrac{\sigma_i}{\psi_i}\right)^{n_i} & \text{for } \sigma_i>0 , \\ 0 & \text{for } \sigma_i \leq 0 , \end{cases} \tag{3.78}$$

where the material constants $A_i>0$, $n_i \geq 1$ can be found by testing appropriate specimens.

The criteria of fracture can be formulated, in general, analogously to those of the isotropic case.

Chapter 4

DAMAGE MODEL FOR DUCTILE FRACTURE

4.1 Stress Field Close to a Slowly Propagating Crack Tip

(a) *Basic Relations.* Consider [25] the problem of a crack $|x| \leq \ell$, $y=0$, Figure 4.1 in a field of creep under longitudinal shear ("antiplane strain"). In this case, the velocities are

$$v_x = v_y = 0 \ , \ v_z = w(x,y) \ .$$

The corresponding strain rates are

$$\dot{\varepsilon}_x = \dot{\varepsilon}_y = \dot{\varepsilon}_z = \dot{\gamma}_{xy} = 0 \ ,$$

$$\dot{\gamma}_{xz} = \frac{\partial w}{\partial x} \ , \ \dot{\gamma}_{yz} = \frac{\partial w}{\partial y} \ . \tag{4.1}$$

In this case we have only the stress components τ_{xz}, τ_{yz}. At Infinity,

$$\tau_{xz} = 0 \ , \ \tau_{yz} = const. = s \ . \tag{4.2}$$

It is assumed that the deformation is small. The strain rate components consist of two parts:

$$\dot{\varepsilon}_{ij} = \dot{\varepsilon}^e_{ij} + \dot{\varepsilon}^c_{ij} \ ,$$

where $\dot{\varepsilon}^e_{ij}$ are elastic components defined by Hooke's law, and the $\dot{\varepsilon}^c_{ij}$ are creep components given by the creep equations (3.1). So, in the case of longitudinal shear, the total strain rates are

$$\dot{\gamma}_{xz} = \frac{1}{G}\frac{\partial \tau_{xz}}{\partial t} + BT^{m-1}\tau_{xz} \ ,$$

$$\dot{\gamma}_{yz}=\frac{1}{G}\frac{\partial\tau_{yz}}{\partial t}+BT^{m-1}\tau_{yz}\ , \tag{4.3}$$

where $B>0$ and $m\geq 1$ are material constants, G is the shear modulus and $T=(\tau_{xz}^2+\tau_{yz}^2)^{1/2}$ is the intensity of shear stresses.

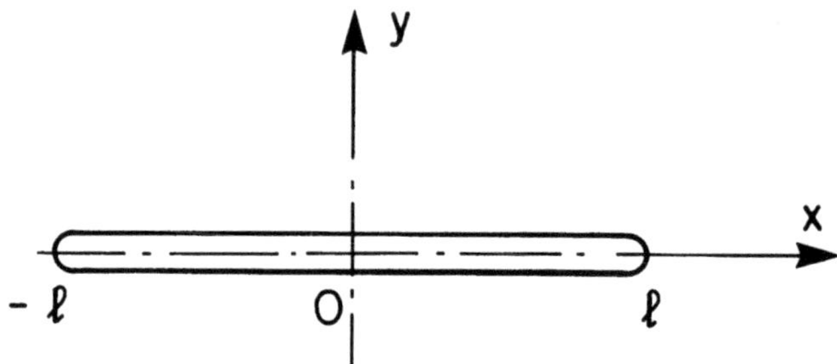

Figure 4.1

Satisfying the differential equation of equilibrium,

$$\frac{\partial\tau_{xz}}{\partial x}+\frac{\partial\tau_{yz}}{\partial y}=0\ ,$$

by the stress function $F(x,y)$

$$\tau_{xz}=\frac{\partial F}{\partial y}\ ,\ \tau_{yz}=-\frac{\partial F}{\partial x}\ ,$$

and substituting the relations (4.3) into the compatibility condition

$$\frac{\partial}{\partial y}\dot{\gamma}_{xz}-\frac{\partial}{\partial x}\dot{\gamma}_{yz}=0\ ,$$

we obtain the differential equation

$$\frac{1}{G}\frac{\partial}{\partial t}\Delta F+BM(F)=0\ , \tag{4.4}$$

where Δ is the Laplace operator

$$\Delta=\frac{\partial^2}{\partial x^2}+\frac{\partial^2}{\partial y^2}\ ,$$

and

$$M(F)=\frac{\partial}{\partial x}[T^{m-1}\frac{\partial F}{\partial x}]+\frac{\partial}{\partial y}[T^{m-1}\frac{\partial F}{\partial y}]\ . \tag{4.5}$$

If the crack does not grow, we have (some time after loading) the steady creep state which is described by the equation

$$M(F)=0\ .$$

Let the crack tip move with the given constant velocity v.

Let us introduce the coordinate system

$$\chi=x-vt\ ,\ y\ \ ,$$

moving together with the crack tip, and the dimensionless variables

$$\xi=\frac{\chi}{\ell_o}\ ,\ \eta=\frac{y}{\ell_o}\ ,\ \tau=\frac{t}{t_1}\ ,$$

where $t_1=\ell_o/v$, and $2\ell_o$ is the initial crack length. The dimensionless stress function is

$$\Phi=\frac{F}{s\ell_o}\ .$$

Then, the differential equation (4.5) becomes

$$\lambda\frac{\partial}{\partial\xi}\Delta\Phi-M(\Phi)=0\ , \tag{4.6}$$

where

$$\lambda=\frac{s}{G}\frac{1}{Bs^m t_1}=\frac{\varepsilon^e}{\varepsilon^c}\ .$$

The parameter λ is equal to the ratio between the elastic strain and the creep strain (accumulated in time t_1) at a long distance from the crack. Differentiation in the operators $\Delta\Phi$, $M(\Phi)$ is carried out with respect to the moving system ξ, η.

(b) *Conclusion.* If the crack tip moves slowly, the parameter λ is small ($\lambda<<1$), the first term in the equation (4.6) can be ignored, and the stress state is described by the equation of steady creep:

$$M(\Phi)=0\ , \tag{4.7}$$

Here, the stress singularity at the crack tip is of the type

$$\rho^{-1/m+1}\ ,\quad (\rho\rightarrow 0)$$

where ρ is the distance from the crack tip. These conclusions result from the analogy between non-linear creep flow and the corresponding power-law of elastic deformation (see [9], [32]).

If the crack tip moves rapidly, the parameter λ is large ($\lambda>>1$), and in equation (4.6) it is possible to ignore the term $M(\Phi)$. Then, the stress field near the crack tip is described by the classical ("elastic") equation

$$\Delta(\Phi)=0\ . \tag{4.8}$$

The singularity is in this case

$$\rho^{-1/2}\ .$$

Since $m>1$, the singularity under steady creep conditions is weaker.

If the index $m>>1$, the stress field can be approximated (in some sense [9]) by the ideally-plastic stress state.

Finally, it can be noted that analogous conclusions are made in the cases of plane strain and plane stress.

4.2 Invariant J_*-Integral in Steady Creep

Let tractions T_i $(i=1,2,3)$ constant in time act on the surface S_T of the body with a crack. If the crack grows slowly, the creep field is steady.

In a plane problem for a non-linearly elastic body there is an invariant J-integral (see, for example, [32])

$$J=\int_C [U dx_2 - X_i \frac{\partial u_i}{\partial x_1} ds]\ ,\ \ (i=1,2)\ . \tag{4.9}$$

This integral has the same value for all paths C surrounding the crack tip, Figure 4.2. The crack runs along the x_1-axis; X_i are the components of the stress vector on the contour C, and

$$U=U(\varepsilon_{mn})=\int \sigma_{mn} d\varepsilon_{mn} \quad (m,n=1,2)$$

is the density of the strain energy; σ_{mn} are stress components, ε_{mn} strain components; u_i are components of displacement. It is well-known that

$$J=-\frac{d\Pi}{d\ell}\ , \tag{4.10}$$

where Π is the total potential energy which is equal to the difference between the strain energy of the body and the work of the external loads.

If the crack length increases, the potential energy Π decreases, and so energy is released. According to Griffith's theory of brittle fracture of an elastic body, the released energy flows into the crack tip where it expends on fracture. In the case of steady creep, as a consequence of the above mentioned analogy, we have the invariant integral.

$$J_*=\int_C [L dx_2 - X_i \frac{\partial v_i}{\partial x_1} ds]\ , \tag{4.11}$$

where

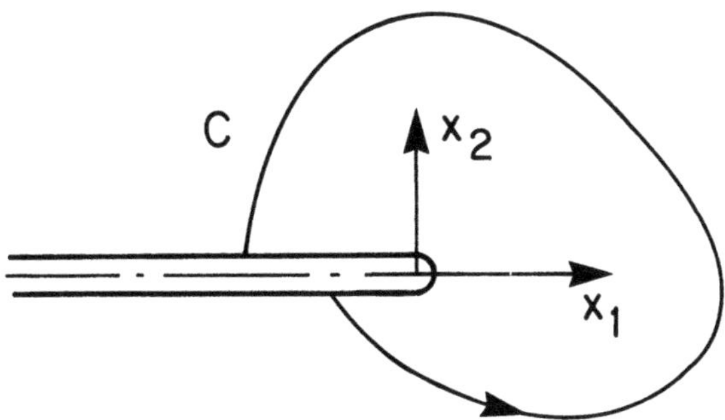

Figure 4.2

$$L = L(\dot{\varepsilon}_{mn}) = \int \sigma_{mn} d\dot{\varepsilon}_{mn}$$

is the potential function proportional to the function of dissipation (section 3.9); v_i are components of velocity.

It is not difficult to prove that

$$J_* = -\frac{dD}{d\ell}, \tag{4.12}$$

where

$$D = \int L dV - \int T_i v_i dS_T \tag{4.13}$$

is the total dissipation rate of the body. The J_*-integral characterizes the intensity of dissipation at the crack tip when the crack extends in length. Thus, the crack tip is the source of dissipation.

The value of J_* depends on the geometry of the body and of the crack, on the boundary conditions, and on the index m in the creep power law.

4.3 Dissipative Criterion of Fracture

The damage of a body under steady creep can be attributed to creep deformation. For simplicity, we assume that the damage does not influence the creep flow.

(a) *Dissipative Criterion of Fracture.* Consider a crack under steady creep conditions. In these problems of crack growth it was suggested by Kachanov [10] and Landes and Bagley [26] that the invariant integral J_* be used.

This suggestion was supported by experimental data.

The time of advance of a unit of length is inverse proportional to the rate $\dot{\ell}$.

The ratio $J_*/\dot{\ell}$ characterizes the accumulation of dissipation at the crack tip.

It is natural to assume that fracture occurs at a certain level of dissipation. This level can be considered as a constant for a specified material. Therefore, we can introduce the dissipative criterion of fracture in the form [10].

$$\dot{\ell}=cJ_* , \tag{4.14}$$

where c is a constant at the given temperature. This constant can be found by testing specimens with cracks under creep conditions.

Note that it is expedient to consider a more general form

$$\dot{\ell}=cJ_*^{\alpha} \tag{4.15}$$

containing two constants c and $\alpha>0$.

(b) *On the Entropy Criterion of Fracture.* A more general entropy criterion of local fracture was proposed by Chudnovsky [28] from the viewpoint of thermodynamics.

In fracture, the entropy density s reaches the critical value s_*, i.e.,

$$s=s_* . \tag{4.16}$$

The critical value s_* may be considered as a "stable" material constant that depends mostly on the type of fracture.

Using the entropy criterion, it is possible to analyze not only mechanical factors but also the physical and chemical ones that affect the process of fracture.

(c) *Crack Growth in a Visco-Elastic Body.* If the process of crack growth is not too slow, it is necessary to take into account the potential energy release resulting from brittle fracture.

Let us consider a linear visco-elastic body of Maxwell's type. Here,

$$\varepsilon_{ij}=\varepsilon_{ij}^{e}+\varepsilon_{ij}^{c} ,$$

where ε_{ij}^{e} and ε_{ij}^{c} are, respectively, elastic and linear creep components of a small deformation.

The body surface S_T is loaded by constant tractions; the crack surface is free.

The stress field in the visco-elastic body coincides with that in the correspnding elastic body (according to the well-known correspondence principle).

The total potential enery associated with the elastic deformation of the body is denoted by Π; the Griffith's fracture work is denoted by W.

Griffith's condition of brittle fracture of an elastic body, i.e.,

$$d\Pi = -dW \,, \tag{4.17}$$

characterizes its limit equilibrium when the crack length ℓ increases by $d\ell$.

In considering a visco-elastic body, we assume that the velocity ℓ of the crack growth consists of the brittle component ℓ_b and the viscous component ℓ_v, i.e.

$$\ell = \ell_b + \ell_v \,. \tag{4.18}$$

According to the relation (4.10),

$$d\Pi = -J d\ell \,. \tag{4.19}$$

The invariant J-integral is proportional to k^2 (k is the stress intensity factor). Griffith-Irvin's condition of brittle fracture can be written in the form

$$J = J_c \,, \tag{4.20}$$

where J_c is the critical value of J. Note that the ratio J/J_c can be interpreted as a driving force when the crack grows by $d\ell$. The energy release rate is determined by the velocity ℓ; according to (4.19) $\dot{\Pi} = -J\ell$. Hence, the brittle component ℓ_b can be considered as proportional to $J\ell$, i.e.,

$$\ell_b = \alpha J \ell \,.$$

where α is a certain constant. The viscous component ℓ_v according to the dissipative criterion of fracture is equal to cJ_*.

Since the stress field in the linear visco-elastic body is identical to that in the elastic body, the integral J_* is proportional to the integral J. Thus,

$$\ell_v = \beta J \,,$$

where β is a certain constant. So, from equation (4.18), we get [37]

$$\ell = \frac{\beta J}{1 - \alpha J} \,. \tag{4.21}$$

The model has two constants α and β according to two different mechanisms of fracture. If $\alpha J = 1$, we have the Griffith-Irvin's condition; then, $J = J_c$ and $\ell = \infty$.

In the case when the material is damaged, we can consider the constants α and β as dependent of the parameter ψ. Note, in conclusion, that a more general scheme leading to a relation of the type (4.21) was given by Chudnovsky, Dunaevsky and Khandogin [38].

4.4 Crack and Damage Growth. Combined Approach

(a) *Influence of Damage on the Crack Growth.* Consider a crack in the field of creep and damage [27]. According to the kinetic equation of damage,

$$\dot{\psi} = -A\left(\frac{\sigma_1}{\psi}\right)^n , \tag{4.22}$$

where σ_1 is the maximal tensile stress in a steady creep field at a given point in the direction of the crack growth.

Equation (4.22) can also be written in the form

$$\dot{\psi} = \psi^{-n}\dot{\psi}_o , \tag{4.23}$$

where $\dot{\psi}_o$ is $\dot{\psi}$ in the undamaged body (at $t=0$). Certainly, equation (4.22) is not applicable at the crack tip. In order to obtain a simple qualitative solution, we use the equations of Hayhurst and Lecki [22] in the following special form:

$$\dot{\varepsilon}_{ij} = \psi^{-n} b T^{m-1} s_{ij} . \tag{4.24}$$

Comparing these relations with the uniaxial relation (2.51), we see that $q=n$, $b=B_1$.

According to experimental data, there is, usually, no significant difference between the values of the indices m and n. In the case of uniaxial tension under constant stress σ and small strain, we find from equations (4.22), (4.23) and the initial condition

$$\psi = 1 , \ \varepsilon = 0 \ \text{ at } t = 0 ,$$

that

$$\psi = \left(1 - \frac{t}{t'}\right)^{1/n+1} , \tag{4.25}$$

where

$$t' = [(n+1)A\sigma^n]^{-1}$$

is the time to brittle fracture. In addition, we have

$$\varepsilon = k(1-\psi) , \tag{4.26}$$

where ε is the creep strain, and

$$k = \frac{B_1}{A}\sigma^{m-n} .$$

Equation (4.26) describes the creep curve including the tertiary portion.

The special form (4.24) of the constitutive equations is convenient for a simple solution (mostly qualitative) of the problems of creep and damage. In this case, the dissipation density rate is

$$d = s_{ij}\dot{\varepsilon}_{ij} = \psi^{-n} d_o \ , \tag{4.27}$$

where d_o is d for the body without damage.

For the damaged body, taking into account equations (4.23), (4.27), and the obvious condition that if $\dot{\psi}\rightarrow 0$, we expect that $\dot{\ell}\rightarrow 0$, we can introduce the generalization of the dissipative criterion of fracture (4.14) in the form [27]

$$\dot{\ell} = c\psi^{-n} J_* \ . \tag{4.28}$$

It is a new criterion of crack growth that cannot be derived from (4.14).

Note that according to experiments by Worswick and Pilkington (Int. Conf. on Fracture, Cannes, France, 1981) the rate of crack growth in the damaged ("pre-cavitated") steel specimens is significantly higher (about on 60%) than the rate in undamaged specimens. The values of ψ and J_* are determined for the steady creep field.

Note also that in the criterion (4.28) it is necessary to understand under ψ the characteristic damage value in the field ahead of the crack tip.

This hypothesis is especially convenient if the stress field ahead of the crack tip can be approximated by an ideally-plastic field. In this field, corresponding to $m\rightarrow\infty$, the stresses are normalized by the value of the external load (see [9]).

Thus, the functions ℓ, ψ are determined by the system of two ordinary non-linear differential equations (4.22) and (4.28), the initial conditions being

$$\ell = \ell_o \ , \ \psi = 1 \ \text{at} \ t = 0 \ . \tag{4.29}$$

When $\dot{\ell}\neq 0$, $\dot{\psi}\neq 0$, the following integrable combination

$$A\sigma_1^n \dot{\ell} = -E\dot{\psi} \tag{4.30}$$

can be obtained, where $E = cJ_*$, $E > 0$. E and σ_1 are, in general, functions of ℓ.

For compressive stress, according to the "dormant damage mechanism" (section 2.3), we assume that $\dot{\psi} = 0$ and $\dot{\ell} = 0$. Now, it becomes possible to analyze low cycle loading.

(b) *Example.* Consider the plane strain problem of a crack ($|x| < \ell$, $y = 0$) growing in the strip $|x| < b$ under tension in the y-direction, Figure 4.3.

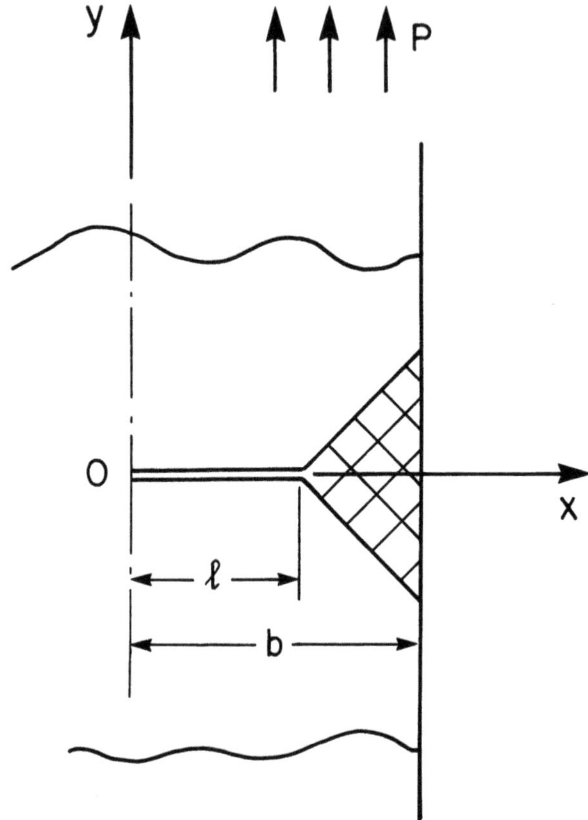

Figure 4.3

According to the solution for the rigid ideally-plastic strip with a crack, we can consider, as an approximation, the triangle of uniform uniaxial tensile stress

$$\sigma_1 = p\left(1 - \frac{\ell}{b}\right)^{-1}, \tag{4.31}$$

at the crack tip, where p is the tensile stress at infinity.

From (4.30), we obtain

$$\psi = 1 - [\Phi(\ell) - \Phi(\ell_o)], \tag{4.32}$$

where

$$\Phi(\ell)=\int_o^\ell \frac{A}{E}\sigma_1^n d\ell\ . \tag{4.33}$$

The expression $\Phi(\ell)-\Phi(\ell_o)$ is a positive monotonically increasing (from zero) function of ℓ, tending to infinity when $\ell \to b$.

Fracture occurs when $\psi=0$. This condition determines the maximal crack length ℓ_*. When $\ell=\ell_*$, $\dot{\ell}=\infty$, and the ligament (ℓ_*,b) is fractured. The corresponding fracture time is

$$t_*=\int_{\ell_o}^{\ell_*} \frac{\psi^n}{E} d\ell\ . \tag{4.34}$$

If the strip is very wide ($b \gg \ell$), we can assume, as a first approximation, that σ_1 $const.=p$. Then, ψ is determined by (4.25). We find from (4.25) that

$$\ell=\ell_o+E\int_0^t \psi^{-n}dt\ . \tag{4.35}$$

The crack length increases gradually and tends to a certain finite length as $t \to t'$, where

$$t'=[(n+1)Ap^n]^{-1}\ .$$

Fracture occurs at the moment t'. In this approximation, the crack growth does not affect the fracture of a very wide strip.

In general, at $x>1$, the further the distance from the tip, the smaller is the stress. Consider a certain fixed point $x' \gg \ell$. The stress at this point is a positive, monotonically increasing function of ℓ, tending to infinity as $\ell \to x'$. Formula (4.32) is applicable for this case, too.

For a wide strip, we can assume that $E=const$. As before, the bracketed expression in (4.32) is a positive monotonically increasing function of ℓ, tending to infinity as $\ell \to x'$. Note that $n \geq m+1$. The condition $\psi=0$ determines the maximal length ℓ_*. When $\ell=\ell_*$, then $\dot{\ell}=\infty$, and the corresponding time t_* can be considered as the time to fracture.

(c) *Conclusion.* The considered model makes it possible to solve various problems of crack growth in the presence of creep and damage under constant or low cycle loading.

It is also possible to consider plane stress problems. But, in this case, it is necessary to take the possibility of "necking" into consideration. Note also that the case of bending of a strip with a crack can be treated in a similar way.

4.5 Dugdale's Crack Model with Developing Damage

Under creep conditions the process of damage accumulation at the crack tip leads to the growth of the crack. The analysis of this non-linear problem presents some difficulties. A simple approach can be made using a modification [35] of the elastic-plastic crack model of Dugdale.

(a) *Constant Load.* Let us consider a crack under tension in a state of plane stress. Let the crack length be 2ℓ, Figure 4.4. At the crack tip there is a small zone of fracture of length a. At high stress, creep is generally accompanied by instantaneous plastic deformation. Therefore, we can assume that the stress at the crack tip is limited by the yield stress σ_y. Thus, we obtain the following model. The stress at the tip zone is equal to the yield stress σ_y. The edges of the middle part of the crack are traction free. Due to the finite stress at the crack tip, we have the well-known relation

$$k_1\sqrt{\pi\ell}-2\sigma_y\ell\cos^{-1}(1-\frac{a}{\ell})=0\,. \tag{4.36}$$

where k_1 is the stress intensity factor.

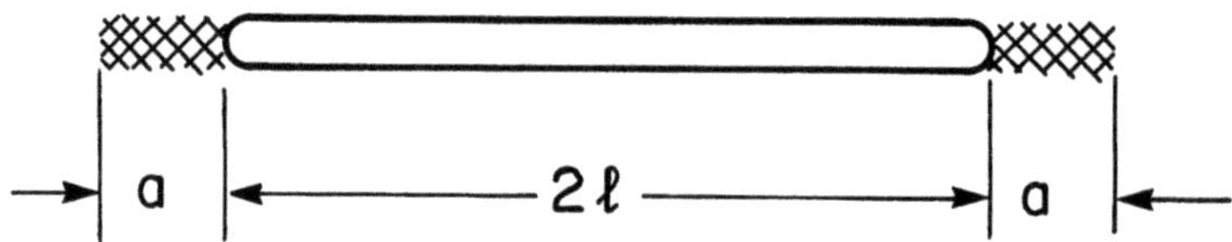

Figure 4.4

In the case of an infinite plate, $k_1=p\sqrt{\pi\ell}$. Hence,

$$\frac{a}{\ell}=1-\cos\frac{\pi p}{2\sigma_y}\equiv K\,. \tag{4.37}$$

Therefore, the length a of the small zone of fracture is equal to

$$a=K\ell\,. \tag{4.38}$$

Next, let us describe the continuity ψ at the tip zone by the kinetic equation (4.9). Then,

$$\frac{d\psi}{dt}=-A(\frac{\sigma_y}{\psi})^n\,.$$

Fracture will take place after the time interval

$$\tau=[A(n+1)\sigma_y^n]^{-1}\,, \tag{4.39}$$

When the tip zone will be fractured, the crack length will increase by the amount $K\ell$, and we obtain again the same problem but with a different length $\ell'=\ell(1+K)$. The process of crack growth will be repeated.

According to this scheme, the crack grows in steps at equal time intervals τ. As the crack length increases, the fracture zone a increases too. Note that a crack advancing in steps has been observed sometimes in tests.

(b) *Cyclic Loading.* Consider, for simplicity, the case of a rectangular cycle. In the first half of the cycle, the stress at infinity is tensile ($p>0$). According to the scheme of "dormant damage mechanism" (section 2.3) damage does not develop when $p<0$. Therefore, fracture of the small zone at the crack tip takes place after the time interval 2τ. Note that Janson and Hult [36] also considered the Dugdale's crack model with damage; they assumed that the damage at the tip zone can be attributed to a certain value of yield strain.

4.6 Approximate Model of Crack Growth

(a) *Introduction.* In simple plane problems, the propagation of a macrocrack can be analyzed by the method described in section 4.4 and is illustrated by the example of the crack in the strip under tension, Figure 4.3. A similar solution is applicable to a strip with two symmetric cracks at the strip sides, since in the middle of the cross-section, the tension is uniform (according to the rigid-ideally-plasic solution). However, in the cases when the crack grows in a non-uniform stress field and does not affect the "limit load", this scheme is not applicable.

A method of approximate analysis of propagation of a macrocrack in a non-uniform stress field will be considered below.

(b) *Basic Relations.* Let the process of damage accumulation become unstable at $\psi=\psi_o$ (note that ψ_o can sometimes be taken equal to zero). As this takes place, one or several macrocracks will appear. If the stress field is uniform, failure occurs as soon as the level ψ_o is reached, because a macrocrack in the field of dispersed damage propagates very rapidly if the field is on the verge of instability. If the stress field is non-uniform, the critical value ψ_o is reached at the moment t_I on a certain part of the body surface where one or several cracks nucleate and propagate into the body. Depending on the ability of a metal to undergo plastic deformation, stresses at the crack tip can be either limited by the yield stress σ_y (plastic behavior) or estimated by the value $k\sigma_1$ (brittle behavior), where σ_1 is the nominal tensile stress in the region at the crack tip and k is a certain correctional coefficient (sometimes called effective coefficient of stress concentration). In the first case, $\sigma_1=\sigma_y$ at the crack tip. It is possible to assume that under creep conditions instantaneous mechanical properties of steels do not significantly depend on the level of damage.

Fracture at the crack tip occurs at $\psi=const.$ Hence,

$$\frac{\partial\psi}{\partial t}+\frac{\partial\psi}{\partial u}\frac{du}{dt}=0\,, \tag{4.40}$$

where u is a distance in the direction of crack propagation. Since the rate of damage is relatively small, we can assume that the damage at a certain distance from the crack tip is approximately characterized by the field ψ_I at time t_I. On the other hand, near the crack, damage rapidly turns to the critical value ψ_o.

Assume that the gradient $\partial\psi/\partial u$ is proportional to the difference $\psi_1-\psi_o$. Then, we obtain from (4.37)

$$\frac{du}{dt}=k(\psi_I-\psi_o)^{-1}\,, \tag{4.41}$$

where k is a constant to be determined experimentally. Since ψ_I is a function of u, equation (4.41) yields

$$\int_o^u(\psi_I-\psi_o)du=k(t-t_I)\,. \tag{4.42}$$

If failure occurs when u reaches a certain value u_*, the time of fracture can be found from (4.42).

The character of the final stage of fracture depends on the ductility of the metal and on the type of loading.

For instance, propagation of a large crack in a ductile metal can lead to a plastic collapse of the body.

Figure 4.5 a) shows the distribution of the continuity ψ_I in the case of pure bending of a beam; $\psi_I=1$ in the zone of compression.

Failure occurs at $u=u_*$ when the limit state is reached. Figure 4.5 b) shows the scheme of plastic flow in the case of plane stress.

Another possible case is that of a relatively small crack that does not affect much the bearing capacity of the body. However, penetrating into the body, a crack can cause leakage which renders the part inoperative (cracks in pipelines, for example).

Consider now the case when the stress at the crack tip is equal to $k\sigma_1$. According to the kinetic equation, $\partial\psi/\partial u$ is proportional to σ_1^n at the crack tip. Using the above mentioned assumption concerning $\partial\psi/\partial u$, we obtain from (4.40)

$$\frac{du}{dt}=k_1\sigma_1^n(\psi_I-\psi_o)^{-1}\,, \tag{4.43}$$

where k_1 is a certain constant. Hence

$$\int_o^u(\psi_I-\psi_o)\sigma_1^{-n}du=k_1(t-t_I)\,. \tag{4.44}$$

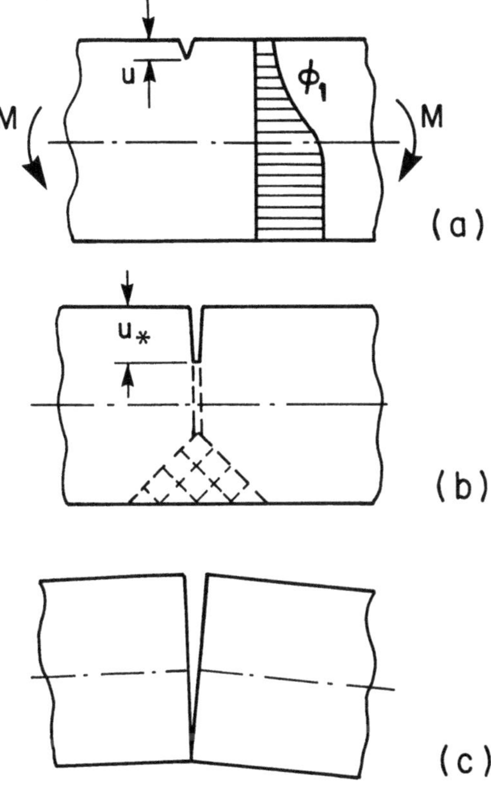

Figure 4.5

Returning to the example of bending of a beam note that the crack propagates with acceleration and cuts the beam into two parts, Figure 4.5 c).

(c) *Local Crack in a Tube under Internal Pressure.* As has been mentioned, in the case of fracture of a tube, loaded by internal pressure, the outer surface of the tube is often covered with a net of cracks that corresponds, to some extent, to the propagation of the fracture front (section 3.5). One of the cracks can propagate through the wall, Figure 3.4, thus rendering the tube inoperative.

As has already been mentioned (section 3.5), if $m>2$, the maximum stress σ_ϕ is reached on the outer surface $r=b_o$ of the tube.

The crack originating there will propagate inside the wall and spread along the generatrix. At a certain moment it will reach the inner surface $r=a_o$.

The tube's wall in the vicinity of the crack originating on the outer surface at the moment t_I is shown in Figure 4.6. The stress σ_ϕ is independent of time for $t<t_I$ and we find from the kinetic equation (4.19)

$$1-\psi^{n+1}=(n+1)A\,\sigma_\phi^n t\ .$$

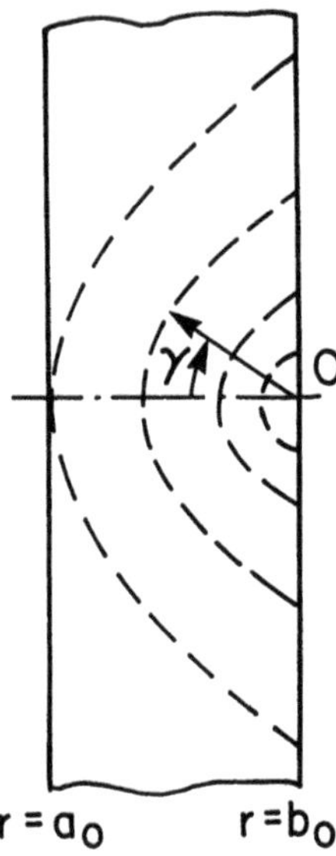

Figure 4.6

The time t_I is found by assuming $\psi=\psi_o$, $r=b_o$. According to this assumption, the field of damage for $t>t_I$ is like that for $t=t_I$, i.e.

$$\psi_1^{n+1}-[1-(1-\psi_o^{n+1})\sigma_\phi^n(r)\sigma_\phi^{-n}(b_o)]\ ,$$

where σ_ϕ is given by equation (3.28) when $b=b_o$.

Let u be a certain direction forming the angle γ with the horizontal; it is evident that $r=b_o-u\cos\gamma$.

According to formula (4.43), the equation of the fracture front is

$$R(b_o-u\cos\gamma)=k(t-t_I)cos\gamma\ ,$$

where

$$R(r)=\int_r^b(\psi_1-\psi_o)dr\ .$$

The crack will reach the inner surface of the tube at the moment $t=t_*$. Assuming $\gamma=0$, $u_*=b_o-a_o$, we find

$$t_*=t_I+\frac{1}{k}R(a_o)\ .$$

The shape of the crack at this moment is determined by the equation

$$R(b_o-u\cos\gamma)=R(a_o)cos\gamma\ .$$

The dashed lines in Figure 4.6 show the successive configurations of the crack; with time they become more "stretched" in the axial direction; initially they are close to a circular arc.

If, at the moment t_I the crack develops along a certain segment of the generatrix, the front of fracture is an envelope of cracks originating from various points of the mentioned segment.

(d) *Concluding Remarks.* The method of the approximate solution suggested above is applicable to some other problems of fracture.

This approach was used by Chrzanowski and Dusza [39] to solve the problem of an edge crack propagation in a strip subjected to either tension or bending.

Chapter 5

DAMAGE MODEL FOR DUCTILE FRACTURE

5.1 Basic Concepts

Large plastic strain in metals is accompanied by nucleation and growth of microvoids and microcracks. This phenomenon is called "ductile plastic damage"; it leads to plastic ("ductile") fracture. This kind of fracture is considered, for example, in the book of McClintock and Argon [1], from the viewpoint of physics. The corresponding continuum damage mechanics model allows a more comprehensive approach to the formulation of a criterion of local fracture.

Considering this model, using the general equations of the continuum mechanics, and applying numerical methods, it is possible to predict the level of damage and the conditions of ductile fracture for different structures.

Below we shall use the simple model of isotropic damage and the concept of actual stress (section 1.2).

The total strain components ε_{ij} consist of two parts: the elastic part ε_{ij}^{e} and the plastic part ε_{ij}^{p}, i.e.

$$\varepsilon_{ij}=\varepsilon_{ij}^{e}+\varepsilon_{ij}^{p} \; . \tag{5.1}$$

Note that in the case of well developed plasticity, the difference between ε_{ij} and ε_{ij}^{p} is insiginficant.

In order to introduce the criteria of damage and fracture, we consider an elastic solid with damage. Then, the density of Helmholtz's free energy f_e at constant temperature is a function of the strains ε_{ij} and the internal variable ψ, i.e.

$$f_e = f_e(\varepsilon_{ij}, \psi) \,.$$

It was shown in section 1.5, that

$$f_e = U\psi$$

where $U = U(\varepsilon_{ij})$ is the density of the elastic strain energy. The generalized force Q associated with the damage parameter is

$$Q = -U \,. \tag{5.2}$$

It can also be written in the form $Q = -\frac{1}{2E}\overline{\sigma}^2$. Here,

$$\overline{\sigma} = T[2(1+\nu) + 3(1-2\nu)(\frac{\sigma}{T})^2]^{1/2} \,, \tag{5.3}$$

where T is the intensity of the shear stresses, and σ is the hydrostatic stress. Thus, $\overline{\sigma}$ is a product of T and a function of the ratio σ/T.

This ratio characterizing the "stiffness" of the triaxial stress state is important for the problems of fracture and damage.

5.2 Damage Criterion

The expression (5.3) was introduced in [31] as an effective stress criterion in damage. But the expression in brackets in (5.3) is quadratic in σ/T and does therefore not depend on the sign of σ.

The value $\overline{\sigma}^2/2E$ is equal to the elastic strain energy. Hence, the criterion of damage in the form (5.3) is analogous, in a sense, to the classical Beltrami energy criterion of fracture; it is known that the Beltrami criterion does not agree with experiments. Generally, it is expedient to modify the expression of the effective damage stress. The simplest way for it is the linear combination

$$\overline{\sigma} = (\alpha T + \beta\sigma) \,, \tag{5.4}$$

where α, β are constants for the given material.

Naturally, some other form $\overline{\sigma} = \overline{\sigma}(T, \sigma, ...)$ (for example, of Mohr's type), free from the above mentioned shortcoming can be introduced. The effective stress $\overline{\sigma}$ as a criterion for damage plays a role analogous to von Mises' criterion T for plasticity.

5.3 Fracture Criterion

At a certain critical value of continuity ψ_c, which is a material constant, the continual damage growth becomes unstable and a macrocrack occurs.

Thus, the criterion of local fracture has the form

$$\psi=\psi_c \,. \tag{5.5}$$

The critical value ψ_c can be found from special (for example, uniaxial) tests.

5.4 Kinetic Equation of Ductile Damage

(a) *Kinetic Equation.* Here, we shall consider only the case of isotropic damage and isotropic plasticity. Then, at a constant temperature, the ductile damage can be described by two variables – the effective stress $\bar{\sigma}$ and the intensity of the plastic shear strain rates

$$\dot{p}=(2\dot{\varepsilon}_{ij}^{p}\dot{\varepsilon}_{ij}^{p})^{1/2} \,. \tag{5.6}$$

Note, again, that for developed plastic deformation metals can be considered as incompressible; it is also possible to consider the total strain rates instead of $\dot{\varepsilon}_{ij}^{p}$.

Since the ductile damage is the consequence of the plastic deformation, the damage rate is a function of $\dot{p}$.

For the case of a power function of the effective stress $\bar{\sigma}$ and a linear function of $\dot{p}$ Lemaitre [31] introduced the following kinetic equation of ductile damage:

$$\dot{\psi}=-A\left(\frac{\bar{\sigma}}{\psi}\right)^{n}\dot{p} \,, \tag{5.7}$$

where $A>0$ and $n\geq 1$ are constants.

(b) *One-Dimensional Case.* In this case, we can take $\bar{\sigma}=\sigma_1$; for an incompressible body,

$$\dot{p}=\sqrt{3}\dot{\varepsilon}_1^{p} \,.$$

Hence,

$$\dot{\psi}=-A\left(\frac{\sigma_1}{\psi}\right)^{n}\sqrt{3}\dot{\varepsilon}_1^{p} \tag{5.8}$$

or

$$d\psi=-A\left(\frac{\sigma_1}{\psi}\right)^{n}\sqrt{3}d\varepsilon_1^{p} \,. \tag{5.9}$$

Let the plastic strain ε_1^{p} be given by a certain law of hardening coupled with damage according to the concept of actual stress, i.e.

$$\varepsilon_1^{p}=f\left(\frac{\sigma_1}{\psi}\right) , \tag{5.10}$$

where f is a monotonically increasing function. Then,

$$\frac{\sigma_1}{\psi}=f^{-1}(\varepsilon_1^p)\ , \tag{5.11}$$

where f^{-1} stands for an inverse function. The variables in equation (5.9) are separated. Integrating, and using the condition

$$\psi=1 \text{ for } \varepsilon_1^p=0\ , \tag{5.12}$$

we obtain the continuity ψ as a function of the plastic strain ε_1^p.

Consider, for example, the case of the Ramberg-Osgood law of hardening, i.e.

$$\varepsilon_1^p=\alpha\left(\frac{\sigma_1}{\psi}\right)^\beta\ , \tag{5.13}$$

where $\alpha>0$, $\beta\geq 1$ are constants.

By calculating, we obtain

$$\psi=1-C(\varepsilon_1^p)^\kappa\ , \tag{5.14}$$

where

$$\kappa=1+\frac{n}{\beta}\ ,\ C=\frac{A\sqrt{3}}{\kappa}\left(\frac{1}{\alpha}\right)^{n/\beta}.$$

If $\dot{\psi}=0$ for $\sigma_1<\sigma_o$, where $\sigma_o>0$ is the damage threshold, the kinetic equation (according to Dufally and Lemaitre [31], [45]) has the form

$$\dot{\psi}=\begin{cases}-A\left(\dfrac{\sigma_1-\sigma_o}{\psi}\right)^n\sqrt{3}\dot{\varepsilon}_1^p & \text{for } \sigma_1>\sigma_o\ ,\\ 0 & \text{for } \sigma_1-\sigma_o\leq 0\ .\end{cases} \tag{5.15}$$

(c) *Three-Dimensional Case.* For this case, we can use the kinetic equation (5.7). The value p is defined by the relation (5.6).

Generalizing the Ramberg-Osgood law in the form

$$p=\alpha\left(\frac{\bar{\sigma}}{\psi}\right)^\beta\ , \tag{5.16}$$

we obtain

$$d\psi=-A\left(\frac{p}{\alpha}\right)^{n/\beta}dp\ . \tag{5.17}$$

Integrating and using the condition

$$\psi=1 \text{ for } p=0\ ,$$

we obtain

$$\psi=1-\frac{1}{\sqrt{3}}Cp^{\kappa}\,. \tag{5.18}$$

The damage threshold $\bar{\sigma}_o$ can be introduced analogously to the one-dimensional case.

Failure occurs when the critical value ψ_c of damage is reached. Note that in some practical cases it is possible to assume that $\psi_c=0$.

(d) *Concluding Remarks.*

1. The scheme considered above is applicable in conditions of simple (proportional) loading.

2. In the engineering problems associated with large strain (for example, in problems of cold forming of metals) it is expedient to use the finite strain theory.

3. More general equations of plasticity with some modifications can be used for constructing the model of ductile damage.

Chapter 6

Fatigue Damage

6.1 Fatigue Damage Accumulation

(a) *Stages of Fatigue Fracture.* In a body subjected to cyclic loading, the process of fracture starts with microcracks nucleating and growing at the initial stage of the process. Then, macrocracks begin to form and propagate, which ends in the fracture of the body.

For a body with considerable initial defects (such as notches, cracks, inclusions) the early ("latent") stage of fracture may be very short or even non-existent. The relation between the length of the latent stage and the length of the stage of macrocrack propagation depends on the geometry of the body and on the nature of defects. The time of crack growth amounts from 10 to 80 percent of the life time of the specimen.

If the body is sufficiently homogeneous, and there is no stress concentration, the latent period may be quite long. The time to fracture in this case can be estimated by using fatigue data for standard smooth specimens.

It is difficult to estimate damage accumulation in the latent stage; it is also difficult, as a rule, to determine the initial moment of the formation of macrocracks.

Therefore, in the problem of fatigue the theory of damage accumulation is usually considered with respect to the entire fracture process without distinguishing the stages of fatigue.

Thus, the damage is defined as

$$\omega = \frac{N}{N_*} , \tag{6.1}$$

where N is the current number of cycles and N_* is the number of cycles to fracture.

It is necessary to distinguish two regimes of cyclic loading:

(1) Low-cycle Fatigue,

(2) High-cycle Fatigue.

The first regime is characterized by high stress level (sometimes higher than the yield stress) and by a relatively low number of cycles to failure (hundreds, thousands).

The high-cycle fatigue is characterized by low stess level and by a large number of cycles to failure (millions).

(b) *Damage Accumulation.* Under the conditions of cyclic loading, the analysis can be generally based on the principle of linear summation of damage. This principle as applied to fatigue fracture was formulated by Palmgren (in 1924) and by Miner (in 1945).

Let $\Delta\sigma_k$ $(k=1,2,...,s)$ be the amplitude of stress in the k-th group of cycles ΔN_k and N'_k the number of cycles to fracture for this amplitude.

According to the principle of linear summation the fracture time (characterized by the number s) is given by the relation

$$\sum_{k=1}^{s} \frac{\Delta N_k}{N'_k} = 1 . \tag{6.2}$$

If the stress amplitude changes continuously, the latter relation takes the form

$$\int_o^{N_*} \frac{dN}{N'} = 1 , \tag{6.3}$$

where N' is the number of cycles to fracture for the current amplitude $d\sigma$; N_* is the number of cycles to failure.

The value

$$\omega = \int_o^N \frac{dN}{N'} \tag{6.4}$$

characterizes the level of damage.

The kinetic equation of damage follows from equation (6.4) as

$$\frac{d\omega}{dN} = \frac{1}{N'} . \tag{6.5}$$

It is obvious that the condition of fracture is

$$\omega=1\,. \tag{6.6}$$

For symmetric cycles of extension-compression (at sufficiently high stress) Manson and Coffin proposed the empirical criterion

$$N_*\Delta_p^2=const. \tag{6.7}$$

Here, Δ_p is the width of the plastic hysteresis loop of the cycle. Hence, the number of cycles to failure depends on the energy dissipated in the cycle.

A more general scheme can be obtained from the kinetic equation (1.1)

$$\frac{d\omega}{d\lambda}=f(\sigma_{ij},\omega,\ldots)\,, \tag{6.8}$$

where $\lambda>0$ is a monotonically increasing parameter similar to time t. It was mentioned above (Chapter 1) that in the case of an irreversible process such a parameter associated with the entropy $s(t)$ can be introduced as

$$\lambda=\int_o^t s(\tau)d\tau\,. \tag{6.9}$$

It should be noted that, instead of $s(t)$, we can consider a certain monotonically increasing function of s.

The choice of the parameter λ and the function f in the equation (6.8) are, as a rule, determined by experimental data.

Now, assume that the fracture process is a result of plastic deformation. Low-cycle fatigue of metals at room temperature and, to some extent, fracture under monotonic loading belong to this case. Then, the rate of loading is insignificant and the Odqvist's parameter

$$\lambda_o=\int(2d\varepsilon_{ij}^p d\varepsilon_{ij}^p)^{1/2}$$

characterizing the "path" of the plastic deformation, can be taken as the parameter λ.

For symmetric cyclic loading,

$$\lambda_o=4N\Delta_p\,.$$

Assuming in the kinetic equation (6.8) that the rate of damage $d\omega/d\lambda$ is proportional to the amplitude of the plastic deformation, and making use of the conditions

$$\begin{aligned}&\omega=0 \text{ at } N=0\,,\\ &\omega=1 \text{ at } N=N_*\,,\end{aligned} \tag{6.10}$$

it is possible to derive (see Novozhilov [30]) the Manson-Coffin criterion (6.7).

For brittle fracture in creep, the "physical" time t can be taken as the parameter λ. This is only the simple version. For a more complicated case, the dissipative creep work

$$\int \sigma d\varepsilon^c$$

where ε^c is the creep strain, can be used as λ.

(c) *Kinetic Equation of Fatigue Damage.* It will be recalled (see Chapter 5) that for ductile plasic damage in the case of strain hardening the increment of damage $d\omega$ is proportional to the stress increment $d\sigma$, i.e.

$$d\omega \sim d\sigma .$$

For example, according to Lemaitre [45],

$$d\omega = [\frac{<\bar{\sigma}-\sigma_o>}{(\sigma_u-\sigma_o)\psi}]^k \frac{<d\bar{\sigma}>}{\sigma_u-\sigma_o} , \tag{6.11}$$

where $\psi = 1-\omega$; σ_u is the ultimate strength, σ_o is the damage threshold, and $\bar{\sigma}$ is the effective stress; k is a certain constant.

Note that the designation

$$<x> = x \text{ if } x>0 \; ; \; <x> = 0 \text{ if } x \leq 0 \tag{6.12}$$

is used in equation (6.11).

If the law of strain hardening is known, it is possible to introduce, instead of $d\bar{\sigma}$, the corresponding increment of plastic strain.

High cycle fatigue can be considered as a cyclic damage process occuring when plastic strain is very small in each cycle. Therefore, it is possible to start from the same type of kinetic equation as for plastic damage, i.e.

$$d\omega = F(\frac{\bar{\sigma}}{\sigma_u \psi}) d\bar{\sigma} , \tag{6.13}$$

where F is a suitable function.

Let the cycle be defined by the maximal value $\bar{\sigma}_{max}$ and the minimal value $\bar{\sigma}_{min}$ of the effective stress $\bar{\sigma}(t)$.

Integration over one cycle gives (Lemaitre [45])

$$\frac{d\omega}{dN} = 2\int_{\bar{\sigma}_{min}}^{\bar{\sigma}_{max}} F(\frac{\bar{\sigma}}{\sigma_u \psi}) d\bar{\sigma} . \tag{6.14}$$

Palmgren-Miner's law of linear summation corresponds to a power function F. In this case,

$$\frac{d\omega}{dN}=\begin{cases}\dfrac{B}{\sigma_u}[\bar{\sigma}_{\max}^b-\bar{\sigma}_{\min}^b] & \text{if } \bar{\sigma}_{\max}>\sigma_f\ , \\ 0 & \text{if } \bar{\sigma}_{\max}\leq\sigma_f\ ,\end{cases} \tag{6.15}$$

where σ_f is the fatigue limit; B and b are constants.

Chaboche's model [4], [31], which involves non-linear summation, corresponds to the kinetic equation

$$\frac{d\omega}{dN}=(1-\psi^{1+\beta})^{\alpha}(\frac{\Delta\bar{\sigma}}{C\psi})^{\beta}\ , \tag{6.16}$$

where α, β, C are constants and

$$\Delta\bar{\sigma}=\bar{\sigma}_{\max}-\bar{\sigma}_{\min}\ .$$

6.2 Creep-Fatigue Damage

(a) *Low-Cycle Fatigue.* In this case, the scheme of "dormant damage mechanism" (section 2.3) can be used.

If ω is the creep damage, then, using the power law, we have

$$\frac{d\omega}{dt}=\begin{cases}A(\dfrac{\bar{\sigma}}{\psi})^n & \text{if } \bar{\sigma}>0\ , \\ 0 & \text{if } \bar{\sigma}\leq 0\ .\end{cases} \tag{6.17}$$

This model leads to the law of linear damage summation; $\psi=1-\omega$.

In order to define the evolution of damage over one cycle it is necessary to perform integration over this specified cycle.

(b) *Creep-Fatigue Interaction.* In the case of high-cycle fatigue under conditions of creep, it is necessary to take into account the effect of creep-fatigue interaction.

In the creep-fatigue model of Lemaitre and Chaboche [17], it is assumed that the total damage ω is the sum of fatigue damage ω_f and creep damage ω_c, i.e.

$$\omega=\omega_c+\omega_f\ . \tag{6.18}$$

For the creep component ω_c, the simple kinetic equation

$$\frac{d\omega_c}{dt}=A(\frac{\bar{\sigma}}{\psi})^n \tag{6.19}$$

can be used. Here, A and n are constants, $\psi=1-\omega$.

Fatigue component ω_f can be described by the equation

$$\frac{d\omega_f}{dN}=(1-\psi^{\beta+1})^{\alpha}(\frac{\Delta\overline{\sigma}}{C\beta})^{\beta} . \tag{6.20}$$

analogous to equation (6.16). Thus,

$$d\omega=A(\frac{\overline{\sigma}}{\psi})^{n}dt+(1-\psi^{\beta+1})^{\alpha}(\frac{\Delta\overline{\sigma}}{C\beta})^{\beta}dN . \tag{6.21}$$

Integration can be performed numerically, the initial condition being

$$\omega=0 \text{ at } t=0 . \tag{6.22}$$

The time to fracture is determined by the condition

$$\omega=\omega_c , \tag{6.23}$$

where ω_c is the critical value of damage.

The experimental data show (see Lemaitre and Plumtree [34]) that the principle of linear summation does not always correctly estimate the time to fracture (extending it). It is possible to better predict it by using the kinetic equation.

6.3 Crack Growth in Fatigue

(a) *Paris' Law.* At the crack tip, the stress field in an elastic body has a singularity of the type

$$\frac{k}{\sqrt{r}} ,$$

where r is the distance from the tip; k is called the stress intensity factor. It depends on both the form of the body and the form of the crack and also on the load.

Under cyclic loading, the stress intensity factor varies over the range

$$\Delta k=k_{\max}-k_{\min} .$$

The analysis of numerous experimental data shows that the increment $\Delta\ell$ of the crack length is determined by the increment Δk.

Paris [41] suggested that

$$\frac{d\ell}{N}=C(\Delta k)^{n} , \tag{6.24}$$

where C and n are certain positive constants. N is the number of cycles. Note that the coefficient C depends on the frequency and on the mean stress of the cycle.

In the case when the crack $|x|\leq\ell$ is in a plate under tension, Figure 6.1, we have

$$k = \sqrt{\pi \ell} p \ , \tag{6.25}$$

where p is the tensile stress at infinity.

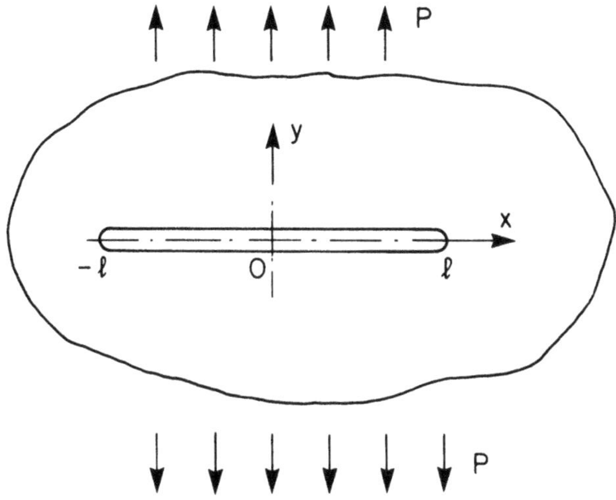

Figure 6.1

Hence,

$$\Delta k = \sqrt{\pi \ell} \Delta p \ .$$

So, equation (6.24) takes the form

$$\frac{d\ell}{dN} = C(\sqrt{\pi}\Delta p)^n \ell^{n/2} \ .$$

Integrating under initial condition

$$\ell = \ell_o \text{ at } N = 0 \ ,$$

we obtain

$$\left(\frac{\ell}{\ell_o}\right)^{\kappa} = \frac{1}{1 - \beta N} \ , \tag{6.26}$$

where

$$\kappa = \left(\frac{n}{2} - 1\right) > 1 \ ,$$

$$\beta = \kappa \ell_o^{\kappa} C(\sqrt{\pi}\Delta p)^n \ .$$

The relation (6.26) is shown in Figure 6.2. When the number of cycles increases in approaching the failure number

$$N_* = \frac{1}{\beta},$$

the rate of crack growth increases fast.

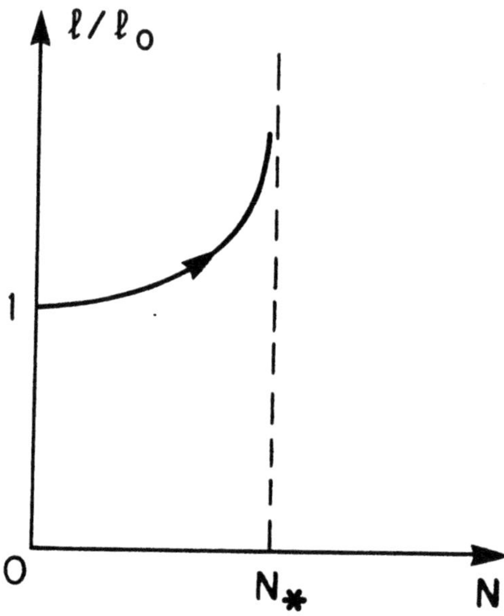

Figure 6.2

REFERENCES

[1] McClintock, F.A. and Argon, A.S., *Mechanical Behavior of Materials,* Addison-Wesley Publ. Co., U.S.A., 1966.

[2] Kestin, J. "Thermodynamics in Thermoplasticity," Lecture Notes for Summer Schooh, Jablonna, Poland, 1973.

[3] Murakami, S. and Ohno, N., "A Continuum Theory of Creep and Creep Damage," in: A.R. Ponter and D.R. Hayhurst (Eds.), *Creep in Structures,* Springer-Verlag, Berlin, 1981, 422-444.

[4] Lemaitre, J. and Chaboche, J., "Aspect phenomenologique de la rupture par endommagement," *J. Mécanique Appliquee*, 2, No. 3, 1978, 317-365.

[5] Cordebois, J. and Sidoroff, F., "Endommagement anisotrope," *J. Mécanique Theoritique et Appliquee*, Special, 1982.

[6] Tamuzh, V. and Lagsdinsh, A. "Variant of Fracture Theory," *J. Mechanics of Polymers*, (in Russian, Riga), No. 4, 1968.

[7] Vakulenko, A., "Superposition in Rheology," *J. Mechanics of Solids, (Mekhanika Tverdogo Tela)*, No. 4, 1970, 69-74.

[8] Kachanov, L.M., "On Creep Rupture Time," Izv. Acad. Nauk SSSR, Otd. Techn. Nauk, No. 8, 1958, 26-31.

[9] Kachanov, L.M., *The Theory of Creep*, Nauka, Moscow, 1960. (In Russian, English translation by Kennedy (Ed.), National Lending Library, Boston Spa, England, 1967).

[10] Kachanov, L.M., *Fundamentals of Fracture Mechanics*, Nauka, Moscow, 1974, (in Russian).

[11] Odquist, F.K.G., *Mathematical Theory of Creep and Creep Rupture*, Clarendon Press, Oxford, 1974.

[12] Rabotnov, Y.N., *Creep Problems in Structural Members*, North-Holland, Amsterdam, 1969.

[13] Krisch, A., "Creep Tests," in: A. Smith and A. Nicolson (Eds.), *Advances in Creep Design*, Appl. Science Publishers, London, 1971.

[14] Staniukovich, A.V., "Brittleness and Plasticity of Heat-Resistance Steels," *Metallurgia*, Moscow, 1967 (in Russian).

[15] Hult, J. and Broberg, H., "Creep Rupture under Cyclic Loading," *Second Bulgarian Congr. on Mechanics, Proc. Vol. 2*, Varna, 1976, 263-272.

[16] Leckie, F.A. and Hayhurst, D.R., "Constitutive Equations for Creep Rupture," *Acta Metallurgica*, Vol. 25, 1977, 1059-1070.

[17] Lemaitre, J. and Chaboche, J., "A Non-Linear Model of Creep-Fatigue Damage Cumulation and Interaction," in: J. Hult (Ed.), *Mechanics of Visco-Elastic Media and Bodies*, Springer-Verlag, Berlin, 1975, 291-300.

[18] Kachanov, L.M., "On Gradual Fracture of Adhesive Bonds," *Int. J. Fracture*, 23, 1983, R143-R146.

[19] Söderquist, B., "Creep Rupture under Uniform Radial Tension of a Disc with a Circular Hole," *Acta Polyt. Scandinavica*, No. 53, 1958.

[20] Chrzanowski, M., "Opis procesu pelzania metali," *Polit. Krakowska Zeszut Naukowy*, No. 3, Krakov, 1973.

[21] Kachanov, L.M., "On Brittle Fracture of a Thin Plastic Interlayer in Creep Conditions," in: G. Dvorak and R. Shield (Eds.), *Mechanics of Material Behavior*, Elsevier Sc. Publ., Amsterdam, 1984, 191-200.

[22] Lecki, F.A. and Hayhurst, D.R., "Creep Rupture in Structures," *Proc. R. Soc. Lond.*, A340, 1974, 323-347.

[23] Hayhurst, D.R. and Leckie, F.A., "The Effect of Creep and Damage Upon the Rupture Time of a Circular Torsion Bar," *J. Mech. Phys. Solids*, 21, 1953, 431-447.

[24] Leckie, F.A., Hayhurst, D.R. and Trampczynski, W.A., "Creep Rupture of Copper and Aluminum under Non-Proportional Loading," University of Illinois, Dept. of Mechanical Engineering Report, Urbana, 1980.

[25] Kachanov, L.M. "Crack Growth in Creep Conditions," *J. Mechanics of Solids, (Mekhanika Tverdogo Tela)*, No. 1, 1978, 97-102.

[26] Landes, J.D. and Begley, J.A., "Fracture Mechanics Approach to Creep Crack Growth," American Soc. for Testing and Materials, STR 590, 1976, 128-148.

[27] Kachanov, L.M., "Crack Growth under Conditions of Creep and Damage," in: A. Ponter and D. Hayhurst (Eds.), *Creep in Structures*, Springer-Verlag, Berlin, 1981, 520-524.

[28] Chudnovsky, A.I., "On Fracture of Solids," in: *Investigations on Elasticity and Plasticity*, No. 9, Leningrad University Press, 1973, 3-41.

[29] Lemaitre, J. Cordebois, J. and Dufailly, J., "Sur le couplage endommagement – elasticite," *C.R. Acad. Sc. Paris*, 288, 1979, B391-394.

[30] Novozhilov, V.V. and Rybakina, O.G., "On Criterion of Fracture," *(Mekhanika Tverdogo Tela)*, No. 5, 1966.

[31] Lemaitre, J. "A Continuous Damage Mechanics Model for Ductile Fracture," *J. of Eng. Materials and Technology*, 107, January, 1985, 83-89.

[32] Hutchinson, J.W., *Non-Linear Fracture Mechanics*, Solid Mechanics, Technical University of Denmark Press, 1979.

[33] Krajcinovic, D., "Creep of Structures – A Continuous Damage Mechanics Approach," *J. Struct. Mechanics*, 11 (1), 1983, 1-11.

[34] Lemaitre, J. and Plumtree, A., "Application of Damage Concepts to Predict Creep – Fatigue Failure," *J. of Eng. Materials and Technology*, 101, July, 1979, 284-292.

[35] Kachanov, L.M., "On Growth of Cracks under Creep Conditions," *Int. J. Fracture*, 1978, R51-R52.

[36] Janson, J. and Hult, J., "Fracture Mechanics and Damage Mechanics a Combined Approach," *J. de Mécanique Appliquee*, 1, No. 1, 1977, 69-84.

[37] Kachanov, L.M., "On Subcritical Crack Growth," *Mech. Res. Comm.*, 3, 1976, 51-52.

[38] Chudnovsky, A.I., Dunaevsky, V.A. and Khandogin, V.A., "On Quasi-Static Growth of Cracks," *Archiwum Mechaniki Stosowanej*, 30, No. 2, Warszawa, 1978, 165-174.

[39] Chrzanowski, M. and Dusza, E., "Creep Crack Propagation in Terms of Continuous Damage Mechanics," in: A. Ponter and D. Hayhurst (Eds.), *Creep in Structures*, Springer-Verlag, Berlin, 1981, 463-474.

[40] Mazars, J. and Walter, D., "Endommagement mecanique du beton," Compte rendu, Laboratoire de Mecanique et Technologie, Universite P. et M. Curie, Paris VI, Cachan, juillet, 1980.

[41]Paris, P. and Erdogan, F., "A Critical Analysis of Crack Propagation Laws," *Trans. ASME, J. Basic Eng.*, No. 4, 1963, 528-534.

[42]Bodner, S.R., "A Procedure for Including Damage in Constitutive Equations for Elastic-Viscoplastic Work-Hardening Materials," Department of Mechanical Engineering Report, University of Illinois at Urbana-Champaign, 1980, 1-15.

[43]Vakulenko, A.A. and Kachanov, M.L.,"Continuum Model of Medium with Cracks," *Mekhanika Tverdogo Tela*, No. 4, 1971, 159-166.

[44]Timoshenko, S. and Woinowsky-Krieger, S., *Theory of Plates and Shells*, McGraw-Hill, New York, 1959.

[45]Lemaitre, J., "Damage Modelling for Prediction of Plastic or Creep Fatigue Failure," Laboratoire de Mécanique et Technologie, Universite P. et M. Curie, Paris VI, Cachan, Rapport No. 3, 1980.

NOTATIONS

$x, y, z; x_i$ $(i=1,2,3)$	cartesian coordinates
u_i	components of displacement
v_i	components of velocity
V	body volume
S	body surface
ν	normal
r, ϕ, z	cylindrical coordinates
θ	temperature
σ_{ij} $(i,j=1,2,3)$	stress components
ε_{ij}	strain components
$\dot{\varepsilon}_{ij}$	strain rate components
T	intensity of shear stresses
H	intensity of shear strain rates
δ_{ij}	Kronecker delta
σ_a	actual stress
$\overline{\sigma}$	effective stress
f	free energy density
s	entropy density
u	internal energy density
U	strain energy density
Π	total potential energy
n, m	constants
ω	damage parameter
$\psi=1-\omega$	continuity parameter

INDEX

GPSR Compliance
The European Union's (EU) General Product Safety Regulation (GPSR) is a set of rules that requires consumer products to be safe and our obligations to ensure this.

If you have any concerns about our products, you can contact us on

ProductSafety@springernature.com

In case Publisher is established outside the EU, the EU authorized representative is:

Springer Nature Customer Service Center GmbH
Europaplatz 3
69115 Heidelberg, Germany

www.ingramcontent.com/pod-product-compliance
Ingram Content Group UK Ltd.
Pitfield, Milton Keynes, MK11 3LW, UK
UKHW021333070726
13610UKWH00011B/63

* 9 7 8 9 4 0 1 7 1 9 5 8 2 *